Environmental Remote Sensing and Systems Analysis

Environmental Remote Sensing and Systems Analysis

Edited by Fred Byron

SYRAWOOD
PUBLISHING HOUSE

New York

Published by Syrawood Publishing House,
750 Third Avenue, 9th Floor,
New York, NY 10017, USA
www.syrawoodpublishinghouse.com

Environmental Remote Sensing and Systems Analysis
Edited by Fred Byron

International Standard Book Number: 978-1-68286-773-0 (Hardback)

Cataloging-in-Publication Data

Environmental remote sensing and systems analysis / edited by Fred Byron.
 p. cm.
Includes bibliographical references and index.
ISBN 978-1-68286-773-0
1. Environmental monitoring--Remote sensing. 2. Remote sensing--Environmental aspects.
3. Earth sciences--Remote sensing. 4. Remote sensing. I. Byron, Fred.
QE33.2.R4 E58 2019
621.367 8--dc23

TABLE OF CONTENTS

PREFACE

Remote sensing is the technique of gathering data and information about an object or phenomenon without making direct contact with the studied entity. This technique has vast environmental applications, which facilitate investigations in the fields of hydrology, geology, glaciology, ecology, geography, etc. Remote sensing is divided into the two classifications of active and passive remote sensing. In active remote sensing, the sensor detects the signal that has been reflected by the observed object. In passive remote sensing, the sensor detects the reflected sunlight. The different data acquisition techniques used for environmental remote sensing are hyperspectral imaging, geodetic remote sensing, radar, LIDAR, etc. This book elucidates new techniques and applications of environmental remote sensing and systems analysis in a multidisciplinary manner. It strives to provide a fair idea of this discipline and to help develop a better understanding of the latest advances within this field. As this field is emerging at a rapid pace, the contents of this book will help the readers understand the modern concepts and applications of the subject.

The researches compiled throughout the book are authentic and of high quality, combining several disciplines and from very diverse regions from around the world. Drawing on the contributions of many researchers from diverse countries, the book's objective is to provide the readers with the latest achievements in the area of research. This book will surely be a source of knowledge to all interested and researching the field.

In the end, I would like to express my deep sense of gratitude to all the authors for meeting the set deadlines in completing and submitting their research chapters. I would also like to thank the publisher for the support offered to us throughout the course of the book. Finally, I extend my sincere thanks to my family for being a constant source of inspiration and encouragement.

Editor

Remote Sensing of Mountain Glaciers and Related Hazards

Pratima Pandey, Alagappan Ramanathan and Gopalan Venkataraman

Additional information is available at the end of the chapter

Abstract

Mountain glaciers are highly sensitive to temperature and precipitation fluctuations and active geomorphic agents in shaping the landforms of glaciated regions which are direct imprints of past glaciations, providing reliable evidence of the evolution of the past Cryosphere and contain important information on climatic variables. But most importantly, glaciers have aroused a lot of concern in terms of glacier area changes, thickness change, mass balance and their consequences on water resources as well as related hazards. The contribution of glacier mass loss to global sea-level rise and increasing number of glacier-related hazards are the most important and current socioeconomic concerns. Therefore, understanding the dynamics of the changes and constant monitoring of glaciers are essential for studying climate, water resource management and hydropower and also to predict and evade glacier-related hazards. The recent advances in the techniques of earth observations have proved as a boon for investigating glaciers and glacier-related hazards. Remote sensing technology enables extraction of glacier parameters such as albedo/reflectance/scattering, glacier area, glacier zones and facies, equilibrium line, glacier thickness, volume, mass balance, velocity and glacier topography. The present chapter explores the prospective of remote sensing technology for understanding and surveying glaciers formed at high, inaccessible mountains and glacier-induced hazards.

Keywords: Mountain glacier, hazard, assessment, remote sensing

1. Introduction

Glaciers require standard and accurate technology to be studied. Remote sensing technologies play tremendous role for monitoring glaciers. In fact, a glacier can be considered as a large body of moving ice wherein water penetrates in the form of snow. The snow then transforms into ice by compaction and recrystallization and the ice flows through the system under its own weight and leaves the system by melting and evaporation [1]. The glacier is thus a large body of moving ice. Therefore, glaciers store considerable amount of fresh water in frozen form. The water supply from the glaciers located in the upstream mountains is vital for sustaining and maintaining downstream cultures. The melt water from the snowpack and glaciers fulfil manifold requirements of humankind. In the dry season, the water from glaciers is released by delayed response through snow and ice melting and enhances the river runoff, therefore providing water to the downstream when there is no other source of water [2]. For instance, the melt water released from the glaciers in the Alps and Himalayas and other mountain ranges is crucially important and plays a major role in the water supply of large downstream population [3–6]. The ice sheets, ice caps and glaciers constitute 10% of the earth's land surface contributing to about 3% of the total water on earth corresponding to about 80% of the world's freshwater [7]. According to the estimate made by Meier and Bahr [8], the total area of the glaciers and ice sheets are about 680,000 km^2 and according to Dyurgerov and Meier [9], the same is about 785,000 km^2. The Hindu-Kush Himalayan region alone contains a total of 60,054 km^2 glaciated area, which is the largest concentration of glaciers outside the polar caps. The Hindu-Kush Himalayan region is home to about 54,252 glaciers and is aptly called as the "Water Tower of Asia" as it provides 86,000,000 m^3 of water annually.

These glaciers feed the world's largest rivers such as the Ganga, Indus, Brahmaputra, Salween, Mekong, Yangtze and Huang Ho and supply water to about one billion people living downstream. The fresh water coming from the glaciers of high mountains in these rivers is an important resource for agriculture, navigation, fishing, generation of hydropower and tourism. Apart from being a boon to society, glaciers also play havoc to life and property of the people residing downstream. The mountain glaciers are a potential source of severe natural hazards [10–12]. Besides playing many roles in hydrological sectors, glaciers are also considered as a key indicator of climate. Any change in the climate is visible through glacier behavior and response. Glaciologists and climatologists carry out research on glacier changes to understand the change in the past and present climate and to predict the future changes. Contribution of glacier melt water to sea level rise under warming climate is the burning topic among the glaciologists and the hydrologists.

Glaciers form under the climatic condition when snowfall is more than snowmelt and this condition in the tropics is fulfilled at very high altitude where the temperature is very less. Therefore, the mountain glaciers are generally located at remote and inaccessible locations. Monitoring of these glaciers through ground survey is costintensive, difficult and sometimes dangerous to life. Remote sensing offers an innovative and valuable tool for gathering information about remotely located glaciers which are otherwise inaccessible and significantly capable of extending the scale of the study both spatially and temporally. In the past few decades, the remote sensing has proved to be a crucial resource for glaciologist. The advent, advancement and increase in the number and quality of earth observing sensors, the development of new technologies, algorithms, high processing capability and new methodologies have brought huge revolution in understanding the Cryospheric processes [13].

Keeping in view the importance of glaciers in society and environment, this chapter will provide information on the remote sensing data available for glaciological studies, the glacier parameters studied by remote sensing and the method of studying those parameters. The chapter will focus on the method of estimating snow and glacier area change, volumetric change, mass change, velocity and assessment of glacier-related hazards. The chapter will broadly address two major topics: (a) study of snow and glacier parameters and (b) hazard assessments. This chapter will provide an overview of the importance, impact and the place of mountain glaciers in our social life as well in scientific research.

The objectives of the chapter are very precise and clear, that is, to endow the readers with the scope of studying various glaciological parts and subjects with remote sensing. Our aim is to make the readers familiar with mountain glaciers, their parts and dynamics and the methodology to study the same. Thus, in the chapter we will attempt to discuss about the remote sensing data types and the different glacier parameters which can be studied and derived by them. The emphasis will be given to the methodology of extracting various glaciological parameters from remotely sensed data. Figure 1 is the field photograph of Chhota Shigri glacier taken during September 2014.

Figure 1. Field photograph of a Himalayan glacier, September 2014 (Chhota Shigri, western Himalaya, India), showing debris on the glacier, the surrounding avalanche prone steep cliffs and the Bergshrund line separating the glacier body from the cliff.

2. Glacier zones and features

Glaciers form when in a year fall of snow is more than the wasting of snow and the trend continues for many years. The formation and sustenance of glacier thus are functions of climatic parameters such as precipitation and temperature. The transformation of snow into glacier ice takes place through compaction and recrystallization [14]. Snowfall, snow avalanches and snow drift are some of the accumulation processes through which glaciers gain in mass, whereas melting, evaporation and calving are the ablation processes by which glaciers lose mass. Climate and topography play major role in determining the shape, size and type of glacier [15]. Starting from the upper elevation to the terminus, a glacier can be divided into several specific zones. A typical temperate mountain glacier consists of (1) accumulation zone, which is the upper most part of the glacier and where there is net gain of ice, and (2) ablation zone, the lower part of the glacier where there is net loss in the ice through melting, calving and evaporation. The accumulation and ablation zones are separated by equilibrium line where there is neither gain nor loss of glacier ice. The lowest part of the glacier where the glacier ends and the discharge starts is known as snout/terminus/glacier toe. A glacier is a dynamic system which along with snow and ice also transports rocks and debris avalanching on the glacier from the side valley walls. These rocks and debris materials are transported through the glacier system from upper zone to the lower zone. Below the equilibrium line, after melting of ice, these rocks and debris concentrates linearly to the sides of the glacier to form lateral moraine. When a tributary glacier meets the main glaciers, the two adjacent lateral moraines form medial moraine. Terminal and end moraines are the rocks and debris piled near the end of the glacier. When these rocks and debris appear on the surface of the glacier through melting of ice, they are called supra glacier debris. Most of the mountain glaciers are debris-covered glaciers. The debris cover on the glacier changes the interaction of glacier with the climate. Sometimes, a glacier ends with a lake near its snout. This type of lake is known as pro-glacier lake. Many times, these proglacial lakes are dammed with moraines. In the enhanced melting condition of glaciers, these lakes can breach the dam and can cause havoc [16].

3. Remote sensing of snow and glacier

The remote sensing is an art and science that can gather information about an object without being in contact with it [13]. The remote sensing system can be airborne or space-borne and uses electromagnetic radiation to collect the information about the object. When the remote sensing system uses naturally occurring radiation, it is called passive remote sensing and when the remote sensing instrument generates its own radiation, it is known as active remote sensing. A glacier surface consists of snow, firn, ice, rock, debris and water, and each component has variable properties in the different electromagnetic spectrum.

3.1. Optical visible and near infrared

The optical visible and near infrared (VNIR) regions of electromagnetic spectrum (0.4–3.0 μm) are the workhorses of remote sensing [17]. The sensors in the VNIR measure radiance radiated from the object, which is related to the reflectance and albedo of the object. Various glaciers zones such as accumulation, ablation, debris covered and water on the glacier have their own specific reflectance characteristics in the VNIR region, based on which the glacier and its various facies can be mapped (Figure 2). Snow has a very high reflectance in the visible wavelength region and a considerable low reflectance in the near-infrared and middle- and short-wave-infrared regions. The reflectivity of freshly fallen snow is very high in visible and infrared regions. Firn, which is one year old snow, has 25–30% less reflectance than snow. The glacier ice has high reflectance in the blue (0.4–0.5 μm) and green (0.5–0.6 μm) wavelength band but sharply decreases to near zero in the red (0.6–0.7 μm) band [17]. The debris on the surface of the glacier significantly lowers the reflectance. The majority of the space-borne sensors operate in number of bands and known as multispectral. One of the most successful, longest and continuous VNIR program is the Landsat program which is continuously observing earth and gathering data since 1972 (Landsat MSS, TM, ETM+, OLI/TIR). The other optical VNIR operating sensors are ASTER, SPOT, MODIS, IRS LISS III/IV and AWiFS, Quickbird and IKONOS. Table 1 lists the spectral regions of optical bands used in Landsat TM and Table 2 presents some of the important satellite missions with their specifications.

Bands	Spectral region
Visible (VIS)	0.45–0.52 (blue)
	0.52–0.60 (green)
	0.63–0.69 (red)
Near infrared (NIR)	0.76–0.90
Short-wave infrared (SWIR)	1.55–2.35
Thermal infrared (TIR)	10.42–12.50

Table 1. The spectral region in different optical bands

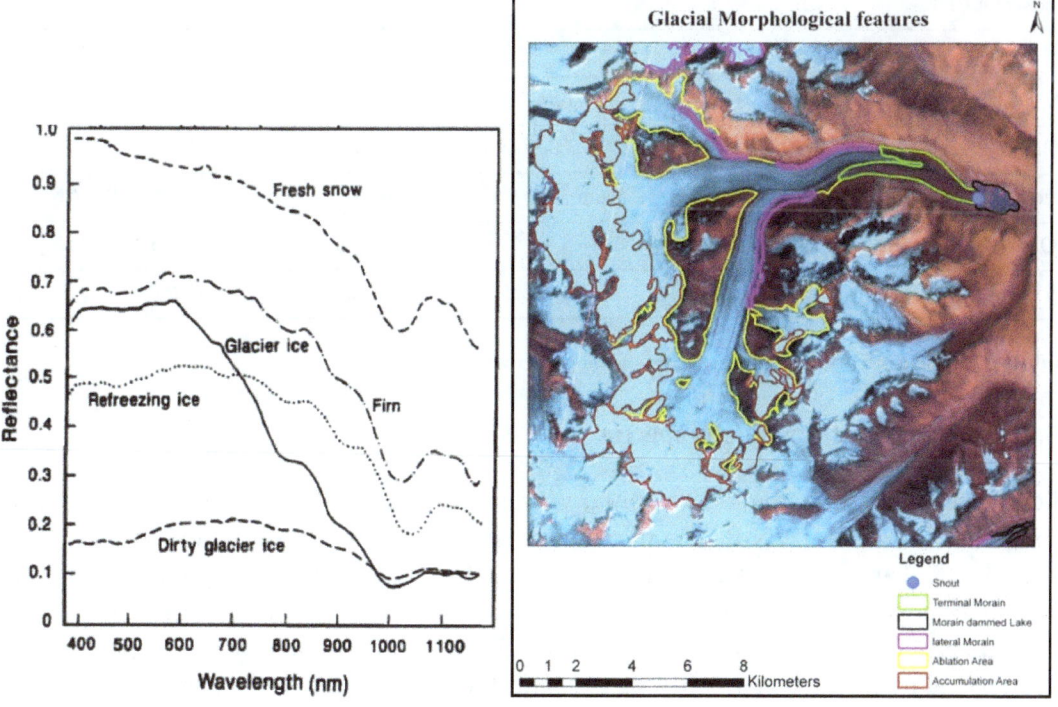

Figure 2. Spectral reflectance curves for snow and ice in different formation stages and satellite image (LISS III, September 11, 2000).

Figure 2 shows the satellite image of Samudra Tapu glacier showing different features of the glacier based on reflectance. As evident from the spectral response curve, the snow has maximum reflectance followed by firn and ice. The debris cover on the glacier has similar reflectance of surrounding rocks. The same can be confirmed from the satellite image of the Samudra Tapu glacier.

3.2. Thermal infrared

The thermal infrared (TIR) (3–15 μm) is a powerful remote sensing tool for discriminating surface objects with different temperature or emissivities [18]. Between the thermal band 8–14 μm, it is possible to measure the temperature of the earth surface and sea surface as atmosphere works as window for these wavelength regions. The surface temperature of glacier is lower than the surroundings and thus can be differentiated using thermal data. The thermally active layer of a glacier has only 10 m depth upto which the seasonal variations can be felt [17]. The most commonly used thermal band sensors for the glaciological study are AVHRR, MODIS, Landsat series and ASTER.

Platform/sensors	Launch	Number of bands	Spatial resolution	Spectral resolution
Landsat MSS	1972		80m	4MS
Landsat TM	1984		15, 30, 60/100m	PAN, 6MS, 1TIR
Landsat ETM+	1999–2003			2TIR, PAN,8MS
Landsat OLI/TIR	2013			
ASTER	1999	15, 30, 90m	14 bands	3VIS/NIR, 6SWIR, 5TIR
SPOT	1984	20m/10m	4 bands	3VIS, 1PAN
MODIS	1999	250, 500, 1000m	36 bands	VIS, TIR
Quick bird	2001	0.6m	4 bands	3VIS/NIR, PAN
IKONOS	1999	1m	4 bands	3VIS/NIR, PAN
IRS LISS III/IV, AWiFS	1988–2011	72 m to 5.8m	4 bands	VIS/NIR

Table 2. List of selected optical remote sensing satellite missions

3.3. Microwave electromagnetic spectra

Microwave spectrum is the most popular wavelength region for studying snow and glacier properties after optical VNIR. The microwave sensors can be passive (radiometer, 3–6 mm spectral range) and active (radar, 1 mm to 1 m spectral range). The atmosphere is transparent in all weather conditions for the whole microwave spectral bands, and therefore, the microwave can be used to study the glacier in all weather conditions and day and night. The major advantage of microwave in monitoring glacier is the ability of microwave signals to penetrate into snow and ice upto various depth and providing information about the internal structure of the glacier. The depth of penetration of the signals depends on the wavelengths. In the dry snow zones, the penetration has been reported to be tens of meters [19]. The L-band radar can be significantly used for collecting information about the glaciers' internal stratigraphy. The ability to penetrate in the wet snow conditions is lesser than the dry snow. With the increase in wavelength, the ability of penetration increases. Surface roughness also influences the reflection and backscattering of microwave significantly. With the usage of synthetic aperture radar (SAR) technology, the spatial resolution of the radar remote sensing can be greatly improved. High-quality and high-resolution SAR data can be used to study glacier facies, glacier stratigraphy and other parameters such as glacier thickness and movement. Table 3 and 4 provide the details of microwave bands and satellites.

Band	Wavelength (cm)	Instrument
Ka	0.8–1.1	–
K	1.1–1.7	–
Ku	1.7–2.4	–
X	2.4–3.8	TerraSAR-X, TanDEM-X, COSMO-SkyMed
C	3.8–7.5	SIR-C, ERS 1/2, ENVISAT ASAR, RADARSAT 1/2
S	7.5–15	ALMAZ
L	15–30	JERS-1, SEASAT, ALOS PALSAR
P	30–100	–

Table 3. Microwave spectrum bands and sensors

System	Country	Year of launch	Band	Resolution (m)
SEASAT	USA	1978	L	25
ERS 1/2	Europe	1991/1995	C	30
J-ERS	Japan	1992	L	18
SIR-C	USA	1994	L	–
X-SAR	Germany/Italy	1994	C/X	15–25
Radarsat-1/2	Canada	1995/2007	C	10–100/3–100
SRTM	USA/Germany/Italy	2000	C/X	90/30
ENVISAT	Europe	2002	C	30, 150, 1000
ALOS	Japan	2006	L	7–100
TerraSAR-X	Germany	2007	X	1–16
TanDEM-X	Germany	2009	X	1–16
COSMOS-SkyMed	Italy	2009	X	1–100

Table 4. List of some selected SAR missions

3.4. Interferometric SAR

Apart from the amplitude of the returned signal, SAR also exploits the phase of the returning signals to extract information of the target. The interferometric SAR (InSAR) technique is based on the phase difference of at least two complex SAR images acquired from either different orbit positions (single pass) or different times (repeat pass). SAR interferometry uses the phase difference between the two returned signals to measure the slight changes in the earth surface. With the single pass interferometry, where the radar is equipped with two antennas, the same point on the ground can be measured at the same time with slightly different angles and this can produce stereo images. These images can be used to produce highly accurate topographic information of the point and can be used to prepare height maps. The InSAR is highly suitable for computing change in the surface thickness of glaciers over large spatial and temporal scales. SRTM is the best example of single pass interferometry, which has been used to produce high-precision global DEM. Tandem data of ERS 1 and 2 (1996/1997) were the first repeat pass SAR data with interferometric generation capability. **T**erraSAR-X **a**dd-**on** for **D**igital **E**levation **M**easurement (TanDEM-X) is the new member of InSAR family along with SRTM, consisting of two satellites TerraSAR-X and TanDEM-X developed by German Aerospace Centre (DLR) and Astrium GmbH. The TanDEM-X (TDX) was launched in June 2010 as an extension of TerraSAR-X in a close formation which enables stereoscopic views. The main aim of this mission was to collect interferometric data over entire global to provide a homogeneous high-resolution global DEM with a relative vertical accuracy of better than 2m within a horizontal resolution of 12 m [20]. The advantage of this single pass bistatic mission is generation of high-quality accurate DEM against low coherence and limited accuracy of data from repeat pass mission. The generation of DEM from InSAR procedure involves interferometry generation, phase unwrapping, multilooking, reflattening, phase to height conversion and geocoding [21]. From the phase difference of returned signals from the two antennas, an interferogram is generated. The phase in an interferogram is influenced by the geometric effects and the topography of the target assuming no movement of the target [22]. By removing the geometric effects, the elevation of a target can be obtained and a DEM can be created [22]. The DEM created by InSAR method is highly accurate and can be used to derive the elevation change of the glacier along with other topographical parameters. The elevation change can further be used to calculate the mass balance of the glaciers. The phase from repeat pass interferometry is the key source for studying small coherent motions of the target between the imaging times. In repeat pass interferometry, the phase difference from the target acquired by the antenna for a nominal time interval enables the measurement of motion of the target during the small acquisition interval. The velocity of the target is obtained by removing the phase obtained due to topography and retaining only the motion phase. The ERS1/2 tandem mission has been extensively used to derive motion of various objects [23]. Figure 3 demonstrates the acquisition geometry of radar interferometry. SAR1 and SAR2 fly on parallel tracks and view the terrain simultaneously from slightly different directions (single pass interferometry) [24]. The technique of InSAR is based on the phase of returned signals from SAR1 and SAR2. The phase difference resulting from the fractional difference of wavelengths of pulse travel time would provide a parallax due to the topography and the shift in location of the target due to motion [13]. The InSAR technique can be exploited to obtain the topographical information and the motion of the target at high precision.

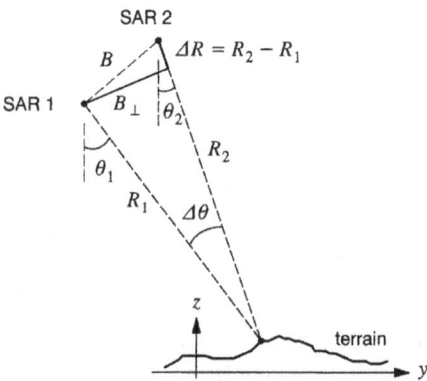

Figure 3. Principal of InSAR acquisition geometry (figure from Balmer and Hartle [24]).

4. Glacier parameters studied with remote sensing

4.1. Snow cover mapping and snowpack properties retrieval

Snow is the most essential and fundamental constituent of a glacier and a key component of earth's energy balance [25]. The mountain snow and the subsequent snow melt can play a dominant role in modulating the local to regional climate and hydrology [26]. The knowledge of snow coverage and snow properties such as albedo, snow grain size, snow depth, snow density and snow water equivalent (SWE) are crucial to know and predict the snow melt. The unique characteristics of snow like high reflectance relative to other surrounding materials (rocks, water, clouds) in the visible part and low reflectance in the mid-infrared part of the spectrum are the foundation of snow cover mapping from space in optical remote sensing [25]. Dozier and others [26] have developed an automatic algorithm to distinguish snow from soil, rocks and clouds by using ratio of reflectance in the VNIR wavelengths (Landsat TM band 2 and 5)which is known as normalized differential snow index (NDSI). According to Dozier [27], a normalized difference snow index (NDSI) is calculated from reflectance in bands at wavelengths where snow is bright (e.g., TM band 2 or MODIS band 1) and where it is dark (e.g., TM band 5 or MODIS band 6), along with a band used for threshold brightness (e.g., TM band 4 or MODIS band 2):

$$\mathrm{NDSI} = \frac{\mathrm{TM\ bands\ 2} - \mathrm{TM\ bands\ 5}}{\mathrm{TM\ bands\ 2} + \mathrm{TM\ bands\ 5}} \tag{1}$$

A snow cover area is mapped when NDSI > 0.4 (Figure 4). Although the snow cover can be mapped with a number of remote sensing devices, multispectral bands in the optical VNIR region of electromagnetic spectrum are most suitable and widely used. In general, the VNIR bands of Landsat MSS, Landsat TM, AVHRR, MODIS, SPOT, ASTER, and IRS have been extensively utilized to map the world's snow cover area. Apart from snow cover mapping, optical VNIR remote sensing has little use in retrieving snow pack properties such as snow depth and SWE. SWE is the most essential snowpack properties in the sense that it represents the total amount of water available if the snowpack has to melt instantaneously [28]. However, in comparison with snow mapping, retrieval of SWE and snow depth through remote sensing has limited success till date; only microwave remote sensing offers measurement of snow depth and SWE as there is penetration through the snowpack at these wavelengths [22]. Most of the studies have used empirical relations to retrieve SWE. The passive microwave radio-meters have been used to retrieve SWE since 1978 [29]. Chang and others [29] have used the difference in brightness temperature at 19 and 37 GHz in the SWE retrieval algorithm to derive SWE from passive microwave. They have used radiative transfer calculation to derive snow depth from SMMR data. The Advanced Microwave Scanning Radiometer–Earth Observing System (AMSR-E) has been used to provide global SWE product since 2002 [30, 31]. The most recent methods of SWE and snow depth retrieval use active microwave data. Rott and others [32] have used Ku-band and X-band for SWE retrieval.

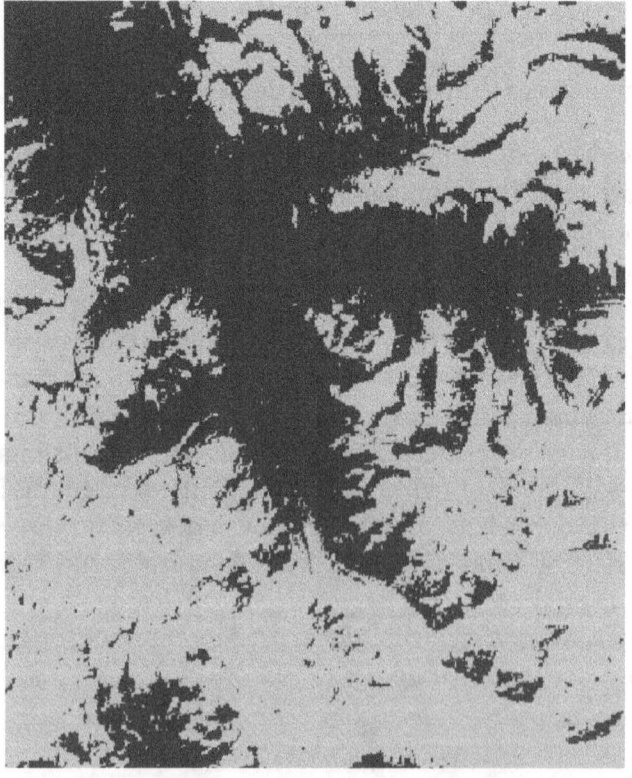

Figure 4. NDSI of Gangotri glacier derived from IRS LISS III image, September 11, 2000, for ablation season. The white part is the snow covered area and the grey is the non snow areas.

4.2. Glacier surface temperature

In global warming situation, the glacier surface temperature is the most important parameter to study the effect of climate change. However, use of traditional method of measuring surface temperature is difficult in mountain terrains. Thermal bands from satellite data provide an excellent alternative for estimating temperatures. TIR can be used to deduce the temperature of snow, ice, clean glacier and debris-covered glaciers. The most widely used TIR sensors are Landsat ETM+, Landsat OLI/TIR, ASTER and MODIS to extract the surface temperature of glaciated areas. In the longer wavelength region, snow acts as a perfect black body. In the infrared region, the strong absorption by snow allows the estimation of temperature from thermal bands [33]. To estimate the surface temperature, the digital number (DN) is converted into radiance. The radiance is converted into surface radiance by reference channel emissivity (RCE) method [33], which then can be converted into surface temperature. The conversion of top-of-the atmosphere (TOA) radiance to surface radiance can be done by following Ref. [34]. Barsi and others [35] have provided a formula to calculate the surface temperature:

$$T_{surface} = \frac{K2}{\ln[K1 + 1/LT]} \qquad (2)$$

where $T_{surface}$ is temperature in Kelvin, K1 and K2 are the calibration constants and L_T is surface radiance calculated as:

$$L_T = L_{TOA} - L_v - \tau(1-\varepsilon) \times L_D / (\tau \times \varepsilon) \qquad (3)$$

where L_{TOA} is TOA radiance, L_v is upwelling spectral radiance between surface and sensor, L_D is downwelling spectral radiance from sky, τ is atmospheric transmittance and ε is the surface emissivity. With the use of thermal band data, it is possible to map the debris-covered glaciers and also mapping of supra and pro glacial lakes can be done with the thermal data. Figure 5 illustrates the thermal map of glaciated region of Chandra-Bhaga basin, Indian Himalaya, using thermal band of Landsat ETM+ for the year 2000.

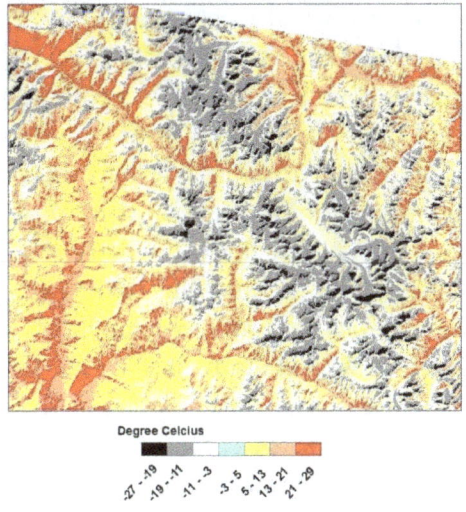

Figure 5. Temperature image of Chandra-Bhaga basin Himachal Himalaya, India, derived from the thermal band of Landsat ETM+ data during ablation season.

4.3. Glacier inventory, monitoring and mapping

Satellite is the backbone of World Glacier Inventory (WGI) and monitoring. Development of new tools and techniques in remote sensing and availability of advanced high-resolution satellite data have brought a revolution in the inventory of world's mountain glaciers. The history of global glacier inventory goes back to 1957–1958, the International Geophysical years, when the inventory of global glaciers was first proposed in the form of national lists of glaciers [35]. This list later known to be WGI under the leadership of Muller and the status was assessed by World Glacier Monitoring Service (WGMS1989), and the digital version of the data was made available by National Snow and Ice Data Center (NSIDC), Boulder, Colorado, USA [35]. A large-scale inventory of global glaciers has been initiated with Global Ice Measurements from Space (GLIMS) in 1995. The GLIMS project was designed to monitor the world's glaciers primarily using data from optical satellite instruments, such as ASTER. GLIMS provide coverage of 58% of global glacierized area with extensive set of attributes [35]. The most recent global inventory is the Randolph Glacier Inventory (RGI), which provides complete collection of digital outlines of global glacierized area excluding ice sheets. The RGI was developed to meet the needs of the fifth assessment of the IPCC on climate change for estimates of past and future mass balance [35]. Satellite images from Landsat 5 TM, Landsat 7 ETM+, ASTER and SPOT 5 HRS have been used to derive the outline of glaciers for RGI.

There are numerous methods to map and delineate mountain glaciers. One of the best methods is the manual delineation of the glaciers through visual interpretation of satellite images acquired during end of ablation season with no recent snowfall. On a false color composite (FCC) image with enhanced contrast, the visual inspection and hence delineation of glacier and other facies become easier. However, to map and monitor glaciers on regional scale, the manual delineation is very cumbersome and time taking and hence not very useful. From the various methods of glacier delineation and mapping on regional scale, the band rationing is simple, robust, accurate, time effective and most suitable [36–39]. The band ratio method is based on the simple rationing of two bands, that is, TM3 and TM5 (RED/SWIR) or TM4 and TM5 (NIR/SWIR) or combination of two bands (NDSI) with application of a threshold [39] with an additional threshold of TM1. The band ratios strongly enhance a specific surface type as well as reduce the bias in illumination from the terrain at the same time. The band ratios method is based on the contrasting response of glacier in the visible and SWIR regions. When the high reflectance of glacier in the VNIR region is divided by the low reflectance in the SWIR region, a high ratio value results [40]. By applying a threshold value, the glacier can be separated from the surrounding rock, soil and vegetation by setting the value above the threshold to black and all others to white (Figure 6). The ratio of TM3/TM5 (RED/SWIR) has the advantage over TM4/TM5 in the sense that it works better in shadows and with thin debris-covered glacier [33, 36, 38]. However, when TM4/TM5 is used, the threshold of TM1 is not required for the fact that TM4 is not much sensitive to atmospheric scattering as TM3 and also rocks in shadows are not mapped [39]. However, Paul and Hendriks [40] have found the NDSI method to be better than the TM3/TM5 method. Andreassen and others [41] have demon-strated the robustness and simplicity of the band ratio method for mapping of glaciers.

Figure 6. Glacier delineation with NDSI showing glaciated (white) and non-glaciated (black) regions, Chandra-Bhaga basin, Himachal Himalaya, India, obtained from IRS LISS III dated September 11, 2000, data.

4.4. Glacier facies mapping with SAR

SAR data can be efficiently used to distinguish the different zones of glacier such as dry snow zone, percolation zone, wet snow zone, firn zone, ablation zone and debris cover. However, the interpretation of SAR data is complicated and difficult than the optical data. The wavelength, polarization, incidence angle, dielectric properties, roughness and grain size are the important glacier parameters that crucially affect the strength of SAR backscatter signals. Based on the contrasting backscatter, the different glacier zones can be mapped with SAR data. Rau and others [42] have identified various radar glacier zones by their backscattering characteristics. These zone are dry snow radar zone, frozen percolation radar zone, wet snow radar zone and bare ice radar zone. Partington [43] has proposed a methodology for facies mapping using multitemporal SAR data. This method involves generation of composite images using winter, early summer and late summer radar backscatter images. Composites are generated by assigning blue to the winter image, green to the late summer image and red to the early summer image. This color combination SAR image is useful to identify different glacier zones due to tonal variations (Figure 7). The color composite will be overlaid on a digital elevation model (DEM). This combination helps to obtain the elevation value as well as backscatter coefficient value for any particular pixel to carry out quantitative analysis. Generally, the winter image defines maximum freezing conditions and late summer image defines maximum melt conditions. This methodology is based on the principle that different zones have typical backscatter signatures related to the snow pack characteristics, influenced by the balance of accumulation and melt at different altitudes.

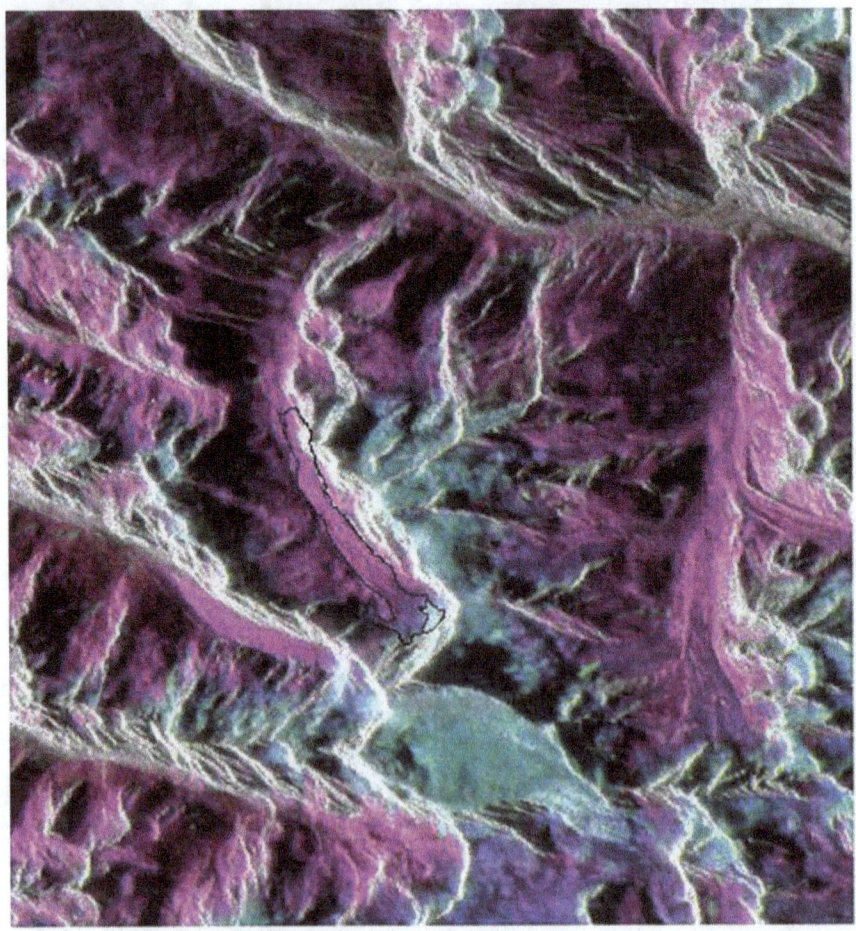

Figure 7. FCC from multitemporal ENVISAT SAR data showing Hamtah glacier region. Three images from winter, early summer and summer have been taken to make the FCC image. In the image, the cyan color indicates fresh snow, blue indicates firn and violet indicates debris cover.

4.5. Equilibrium line altitude extraction

The equilibrium line elevation can be extracted by overlaying an optical image on a DEM. The snowline altitude at the end of ablation season is supposed to be coinciding with the equilibrium line altitude. The cloud-free VNIR optical images with no recent snowfall of ablation season are selected for this purpose. To demarcate ELA on glacier and to differentiate snow from firn and ice, reflectance images are classified (supervised classification by giving maximum number of training classes for better accuracy). The classified images are then draped over a DEM (SRTM has been proved to be good for the purpose) to get elevation points. Before extracting the elevation points, the DEM and the optical images are brought to a common platform in terms of resolution and datum, and hence they are required to be reprojected, resampled and co-registered properly with each other. The classified images are draped on the DEM, and elevation points are determined along the demarcated line for the glacier. The average of the elevations along the line is considered as the ELA of the glacier (Figure 8).

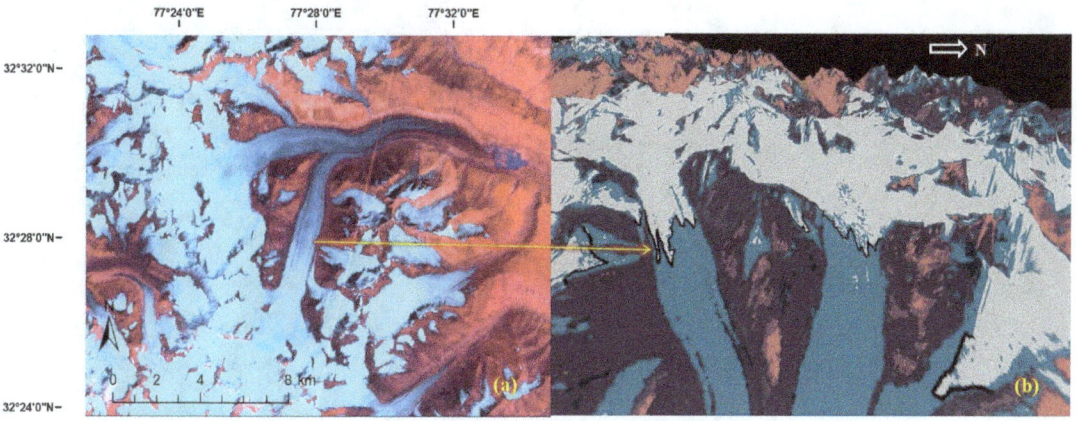

Figure 8. Method of ELA extraction on image: (*a*) reflectance image from IRS LISS III for Samudra Tapu glacier; (*b*) the demarcated ELA on the classified LISS III image dated September 11, 2000, draped on SRTM DEM.

4.6. Glacier topography and morphometry

If climate is the driving force behind the glacier change, the glacier topographical parameters are the controlling factors that modulate the changes. Glacial topography is an important factor that explains the variability in the recessional rates of glaciers of the same basin [44]. The topographical parameters of a glacier can be listed as maximum, minimum, median and mean elevation of the glacier, the altitude range of the glacier, slope and orientation of the glacier. Derivation of topographical parameters of the glacier requires DEMs. Properly co-registered visible optical image overlaid on a DEM can be used to extract the maximum elevation of the glacier, the elevation of snout and equilibrium line altitude. SRTM and ASTER GDEM are the two freely available global DEM which have been extensively used to derive the topographical parameters of the glaciers along with the used of images from Landsat series, ASTER, SPOT and IRS series. The average slope and mean orientation of the glacier can be extracted from the SRTM or ASTER GDEM in ArcGIS (Figure 9). The compactness ratio, the relative upslope area and the slope of the upslope area are the glacier indices which provide the information about the contribution of the avalanching from the surrounding to the glacier and affect the mass balance of the glacier. The method of calculation of these glacier indices has been discussed in Refs [45, 46]. The compactness ratio is the measure of glacier morphometry and can be derived from the formula $(4\pi area)/(perimeter)^2$ following Refs [45, 46]. The relative upslope area is defined as the ratio of the upslope area to glacier surface area and represents the contribution of the surrounding upslope area in the glacier mass balance. The upslope area and the mean slope of the upslope area of glaciers are calculated from the optical data in the visible region along with a DEM in the ArcGIS environment.

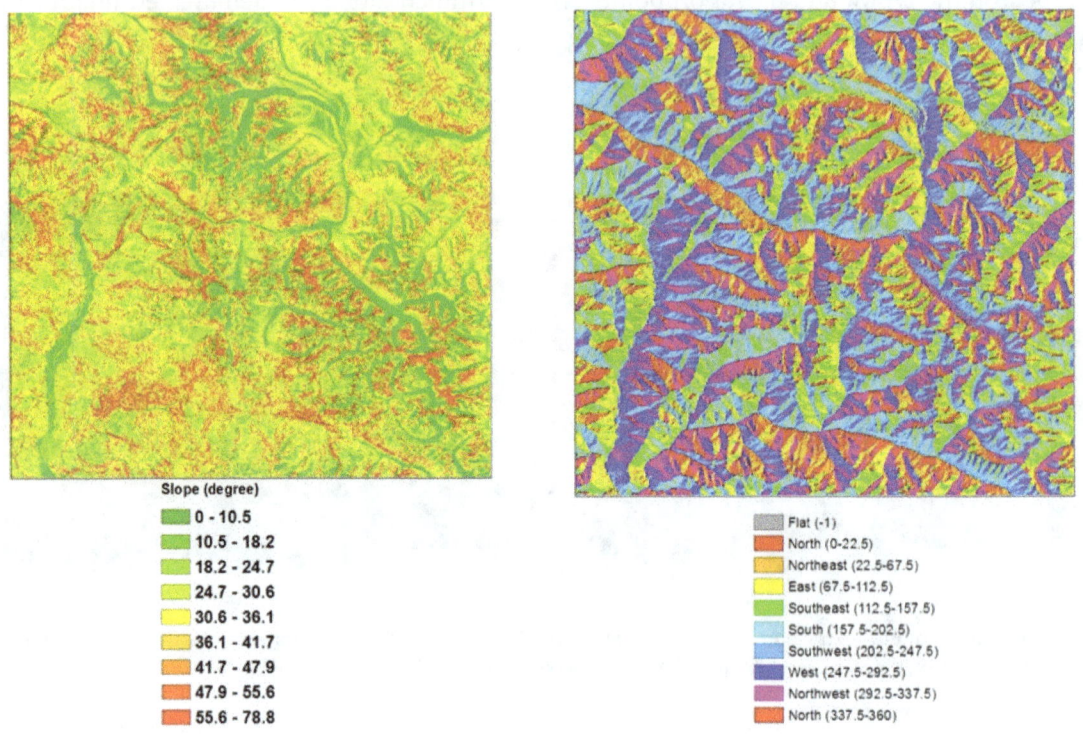

Slope (degree)

- 0 - 10.5
- 10.5 - 18.2
- 18.2 - 24.7
- 24.7 - 30.6
- 30.6 - 36.1
- 36.1 - 41.7
- 41.7 - 47.9
- 47.9 - 55.6
- 55.6 - 78.8

- Flat (-1)
- North (0-22.5)
- Northeast (22.5-67.5)
- East (67.5-112.5)
- Southeast (112.5-157.5)
- South (157.5-202.5)
- Southwest (202.5-247.5)
- West (247.5-292.5)
- Northwest (292.5-337.5)
- North (337.5-360)

Figure 9. Slope and aspect map derived from SRTM DEM for Chandra-Bhaga basin, Indian Himalaya.

4.7. Glacier thickness and mass balance

Glacier mass balance is the most important glacier parameter to be measured and is of interest to glaciologist, climatologist and hydrologists. In a hydrological year, the net gain or loss of the glacier mass is known as glacier mass balance. The glacier mass balance is direct, un-delayed and un-filtered response of climate. Mass balance of glaciers reflects the precipitation and temperature conditions surrounding the glacier and hence is studied to infer the condition and/or variability of climate. Due to the remote location, vastness and irrepressible nature of the Himalayan glaciers, remote sensing-based techniques offer effective alternatives to field-based measurement of mass balance of glaciers. The direct/glaciological surveys of glaciers for mass balance is not feasible for a large number of glaciers as many glaciers does not fulfill the criteria of benchmark glaciers in terms of size, length, geometry, altitudinal range, accessibility and safety. Geodetic mass balance measurement derived from elevation comparisons method complements glaciological method for large number of glaciers. In this method, the change in surface elevation of glaciers is derived by differencing two DEMs of different times. The brief

methodology of deriving glacier mass change from DEMs has been illustrated by methodological chart in Figure 10.

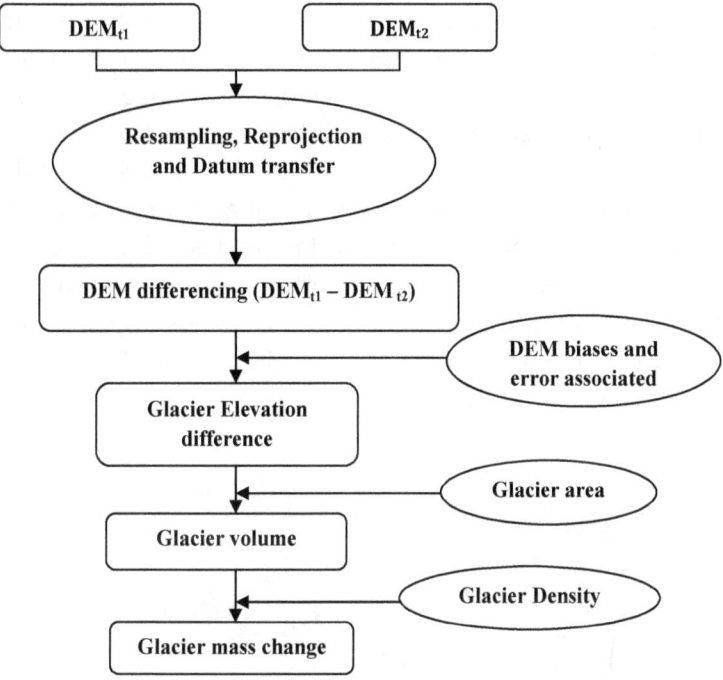

Figure 10. Flow chart showing the methodology of estimating geodetic mass balance of glaciers.

The change in elevation is converted into volume change by multiplying the surface thickness
change with the area of the glacier. Now using the density of glacier, the change in volume is
converted into mass change.

$$dv_z = A\sum dh_z \tag{4}$$

$$\frac{dv}{dt} = A\sum_1^z \frac{dh}{dt} \tag{5}$$

$$\frac{dm}{dt} = \rho\sum_1^z \frac{dv}{dt} \tag{6}$$

where dv_z is volume, dh_z is the elevation change curve, dm is the mass change, A is area and ρ
is glacier density. The estimation of mass balance through elevation comparisons method has
become frequent with the increasing number of available elevation measurements from
satellites data such as ICESat, TanDEM-X, SPOT5 and SRTM and aircrafts [47, 48]. The geodetic

mass balance has been found to be more accurate for longer periods [49] and has also been used to correct the biases in the in-situ direct measurement [50, 51]. Besides, due to the ability of large spatial coverage of satellite data, the method is able to determine mass balance on regional scale [39, 48]. However, the most vital assumption in converting mass change from elevation change is the density of snow/ice lost or gained [52, 53]. In geodetic method, glacier surface elevation is converted into volume change and with the knowledge of density of material lost or gained; the volume is converted into mass change [54]. It is assumed that the density profile remains unchanged and only ice is lost or gained from glacier surface [14, 51]. The assumption of glacier density is taken from Sorge's law, which states that "the density of snow at a given depth below the surface does not change with time" given rates of melting near the surface and refreezing at depth are constant and equal. It follows from Sorge's law that a change of glacier thickness can be converted to an equivalent change of mass by multiplying by the density of glacier ice [47]. Figure 11 shows the elevation change map of a glacier in Chandra basin derived by subtracting SRTM of the year 2000 from TanDEM-X DEM of the year 2011.

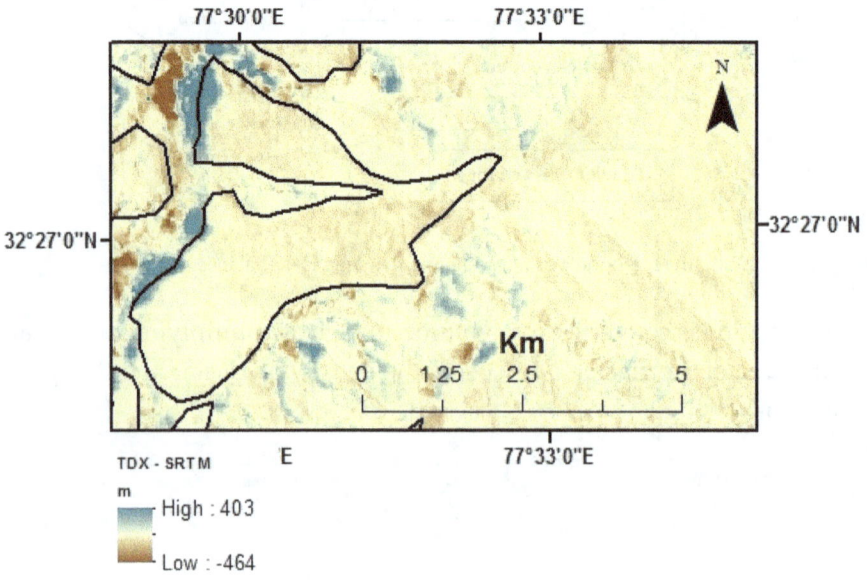

Figure 11. Elevation change map of a glacier of Chandra basin, Himachal Himalaya, India, derived by subtracting SRTM (2000) from TanDEM-X DEM (2011).

4.8. AAR method of mass balance estimation

The mass balance of mountain glaciers at regional scale can be inferred from accumulation–area ratio (AAR) and ELA derived from satellite data. The method is discussed elaborately by Kulkarni [55], who has used this to derive the mass balance of Himalayan glaciers using AAR method on basin scale. The method is based on the relation between AAR/ELA and mass balance of glacier. The AAR is the ratio between the accumulation area and the area of an entire glacier [56]. The AAR of a glacier is characteristic of glacier mass balance and also indicates the state of health of glacier. AAR of a glacier is closely linked with its mass balance. The variation in the AAR of a glacier from year to year can be used as an indicator of variation in net mass balance [14]. Since it is practically not feasible to monitor large number of glaciers on field, hence even the mass balance data of benchmark glaciers are not available for long time series. AAR method has been used as an alternative to estimate the mass balance of glaciers at many regions [55]. This method involves establishing a relationship between AAR and specific mass balance from long-term field observation. A regression equation is constructed between AAR and mass balance with AAR on the x-axis and specific mass balance on the y-axis. The equation obtained is then used to derive mass balance by using AAR values estimated from remote sensing. The mass balance of glaciers can be estimated through this method by using remote sensing data for the periods during which field data are not available. AAR can easily be determined from satellite images. Landsat data, IRS data, ASTER data at medium-resolution scale and SPOT, Quickbird and IKONOS data at higher scale can be utilized to obtain AAR at high precision. To determine the AAR, images at the end of ablation season without cloud cover and recent snowfall are required. The accumulation area can be easily determined by differentiating accumulation zone from ablation zone either manually or by various classification methods. The division of accumulation area from the total glacier area will give the AAR. In Figure 12, an equation has been developed from the linear relation between the specific mass balance and the AAR of Chhota Shigri glacier from the field. From the relation, the following linear equation has been obtained:

$$y = 0.038x - 2.455 \tag{7}$$

In this equation, the x is AAR of the glacier and y is specific mass balance. If we derive AAR from remote sensing data, from the above equation we can compute the specific mass balance of the glacier.

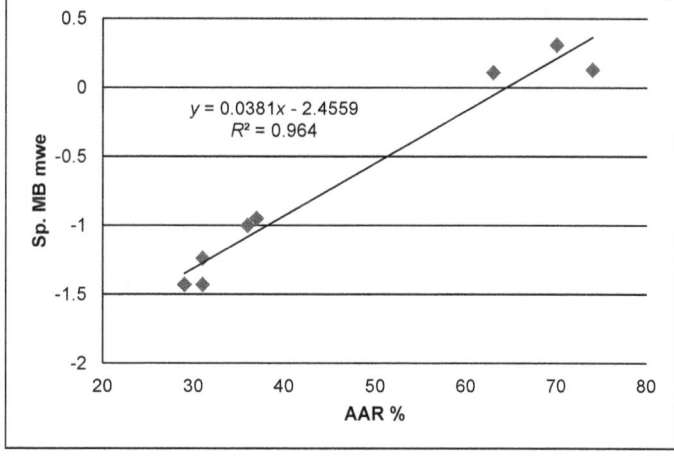

Figure 12. Example of relationship between specific mass balance and AAR, established from field data of Chhota Shigri glacier (data from Ramanathan 2011).

4.9. Glacier velocity

4.9.1. Glacier velocity with feature tracking

Study of glacier velocity provides an understanding of various ongoing dynamical processes of the glacier such as ice flow and ice instabilities, ice flux, mass transportation, development of surge and also the formation and growth of glacier lakes and associated hazards [37, 39]. As the global temperature is reported and predicted to be rising, the glaciers on average are experiencing negative mass balance. In response to the negative mass balance, glacier surface velocity is found to be slowing down in mountainous regions [57, 58]. The glacier surface velocity and movement can be tracked by both optical and SAR satellite data on regional scale. With sequential satellite imageries, the glacier velocity can be determined by tracking glacier surface features such as crevasses and big boulders. This method of calculating glacier velocity with repeat optical and SAR satellite data is known as feature tracking method in general, image matching in optical domain and offset tracking in microwave domain [39]. The temporal baseline in the optical domain can range from weeks to years whereas in the microwave domain it is within weeks. The key point of image matching is the precise co-registration and cross-correlation of the two repeat pass images. Also, the temporal baseline of the repeat images should be such that the displacement of the glacier should not be larger than the accuracy of the method and surface changes due to melting, snowfall and deformation should be very small so that the intensity can be matched properly [39].

A correlation matching is commonly used to obtain both azimuth and range-direction offsets based on intensity pattern patches of two repeat-pass SAR image acquisitions. Through oversampling of the correlation surface, the matching peak can be determined to a small fraction of a pixel. The range offset and the azimuth offset are detected from cross-correlation matching. The successful estimation of the local image offsets depends on the presence of nearly identical features in the two SAR images at the scale of the employed patches. If coherence is retained, the speckle pattern of the two images is correlated and tracking with small image patches can be performed to remarkable accuracy.

The most popular and widely used optical repeat pass satellite images to determine glacier velocity are from Landsat TM, Landsat ETM+ pan, ASTER and SPOT [59, 60]. In the microwave realm, the Envisat ASAR, ALOS PALSAR and TerraSAR-X have been used for offset tracking to calculate glacier surface flow.

In the feature matching technique, the repeat images are co-registered by cross-correlation applied on stable nonmoving areas. Glacier features such as crevasses or debris and big rock boulders which are detectable in images are generally preferred for tracking [61]. The glacier velocity can be determined from the temporal separation and the surface displacement. In recent years with the advent sophisticated computer software and tools as well as high precision remote sensing data, many glaciologists have determined the glacier flow velocity successfully with high accuracy. Luckman and others, Quiney and others and Rankl and others [61–63] have shown that the technique of feature/offset tracking is suitable for Himalayan glaciers due to the presence of respective features.

In the example shown in Figure 13, SAR intensity tracking technique is used for glacier 2-D velocity estimation. The TerraSAR-X high-resolution spotlight mode images acquired on September 27 and October 8, 2012, are used. These images are acquired over the Gangotri glacier, Uttarakhand, India. The estimated surface velocity is varying from 0.1 to 1.1 cm/day over glaciated area (along the medial axis from the accumulation zone to the snout).

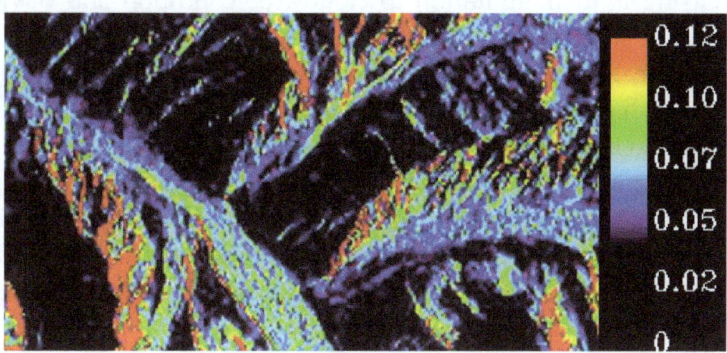

Figure 13. Gangotri glacier velocity estimated using offset tracking method employing TerraSAR-X images. The velocity values are in cm/day (figure and results provided by M. Surendar, CSRE, IIT Powai).

4.9.2. Glacier velocity using SAR interferometry

Goldstein and others [64] for the first time have determined the glacier surface velocity from InSAR data. In InSAR technique, the phase information of radar acquisition from two receiving antennas, separated in either time (repeat track) or space (single track), is used. Two SAR images will have a different distance from target when they are taken from an orbit separated by temporally/spatially from each other. An interferogram can be generated by subtracting the phase of the two images, the phases of which contain range difference. When there is no motion of the target, the phase is influenced by topographical and geometrical effects. If the geometrical effect is removed, the topographical information can be extracted from the phases in interferogram. Now, if the target is moving, then having removed the geometrical effect and the topographical effect, the motion of the target can be measured from the interferogram. ERS tandem data and TerraSAR-X data have been used widely to find the surface velocity of mountain glaciers as well as ice sheets by InSAR method.

5. Comparison between optical, thermal and microwave for Cryospheric studies

The optical remote sensing is based on the detection of reflected solar radiation from surface of the earth in VNIR regions of electromagnetic spectrum which range from 0.4 to 2.5 µm. The basis of TIR remote is the emitted radiation in the spectral range between 8 and 14 µm. Glacier surface has unique spectral properties in the visible-infrared and thermal region, which makes it possible to identify and monitor by optical remote sensing sensors. Optical data at high and medium spatial resolutions from SPOT5, IKONOS, Quickbird, Landsat TM, ETM+, IRC-1C and ASTER are highly useful for regular monitoring and mapping of glacier. The optical remote sensing is exceedingly of use for temporal change analysis of spatial extent of snow and glacier area. Aided with a DEM, information about the glacier geometry and topography can be obtained from optical data. The thermal bands from satellite data have the potential to distinguish debris cover on the glacier. However, one of the main drawbacks of working in optical remote sensing is their limitation to cloud-free condition and daylight, which are sometimes not possible in mountain region where there are always possibilities of forming cloud due to orographical effects. The active microwave system has the capability of acquiring data at all weather conditions, during any time of the day. The microwave remote sensing is more effectively used in extracting snow properties such as snow depth, snow wetness and SWE and glacier facies mapping. The emerging technologies of InSAR and DInSAR have great potential in deriving glacier volume change, mass balance, surface elevation change and glacier velocity. Thus, for studying the evolution and dynamics of mountain glaciers, the complemented usage of optical, thermal and microwave remote sensing is needed.

6. Glacier hazards

The strong interrelation between climatic changes, glacier recession and increasing number of glacier-related hazards is evident in many mountainous parts of the world including the Alps and the Himalayas [65, 66]. Fundamental changes are taking place rapidly in the high mountain regions due to continued global warming [65]. In consequence to rising temperature and climate change, it is predicted that the existing glaciated may soon transform into new landscape with vegetation sparse-bare lands, loose debris and abundant lakes [65]. Such newly transformed landscape definitely would not be in equilibrium with the ecosystem and would thus cause many hazards in order to balance with the system. The most dangerous glacier-related hazards are formation and growth glacier lakes, glacier lake outburst floods (GLOFs), debris and mud flow triggered by flood, snow/ice/rock avalanches and development of crevasses which pose threat to both life and livelihood and brings devastation to mankind and infrastructure including hydropower [67, 68]. The glacier-related hazards has the potential to cause huge casualties in one single event, the damage amounting to hundreds of million [69]. Thus, the risk of loss of life and the devastation of infrastructure are the main motive for studying glacier-related hazards. Monitoring, assessment and management of glacier-related hazards are highly required for the timely prediction of catastrophes and saving of lives downstream. However, due to remote location, complicated terrain, harsh environment and political restrictions, it is not possible to monitor the mountain glacier-related hazards by field observations. The launching of high-resolution satellites in recent decades, emergence of sensor technologies and development of sophisticated tools have posed remote sensing as effective and efficient alternatives to monitor, assess and manage the mountain glacier-related hazards. The optical spectral region of remote sensing is most suitable for glacier hazards assessment. The nature, characteristics, size and growth of hazards decide the selection of remote sensing data. Fusion of multispectral data with the DEMs is the most promising method of glacier hazards monitoring and assessment. The medium-resolution data from Landsat TM/

ETM+/OLI/TIR, ASTER and IRS LISS III can cover regional- to global-scale hazard assessment, whereas high-resolution data such as Quickbird and IKONOS can contribute in providing detailed information [70]. The geometry of the potential dangerous hazardous sites can be obtained from ASTER DEM, SRTM DEM and DEM from other sources. The geometrical assessment of mountain terrain with the help of DEM can provide information about the potential sites of hazards.

6.1. Glacial lake outburst floods

In response to warming of climate, the increasing number and volume of glacier lakes are raising wide concern. Regular monitoring of supra and pro glacial lakes are the key parameter to identify the glacier lake hazards [71–73]. Most of the glacier lakes form near the snout of the glacier and are dammed by unstable moraines and are called moraine dammed lakes. The enhanced melting of glaciers due to rising temperature amplifies the storage of water in the lakes. This occasionally may lead to the breaching of the moraine dams, releasing huge amount of lake water, which in its course gathers the surrounding debris along with it and cause destruction in the downstream. This phenomenon of flash flood is known as GLOF and is one of the most severe catastrophes to occur in the Alpine and Himalayan regions. Richardson and Reynolds [66] have suggested three mechanism of glacier outburst: the rupture of an internal water pocket, the progressive enlargement of internal drainage channels and catastrophic glacier buoyancy. The term GLOFs is most commonly used for the glacier flash floods of Himalaya. A large number of GLOFs have been recorded in central and eastern Himalaya [67, 74]. Compared to the central and eastern part, the western Himalayas have seen lesser number of GLOFs. The application of modern remote sensing technology to locate and monitor the formation and growth of potentially dangerous lakes is necessary due to their far reach. The glacier dynamics, probability of formation and future development of lakes can be assessed by time series of multispectral images. DEMs are found to be crucial in the assessment of moraine dam characteristics, dam geometry, surface material and geometry. The visual interpretation of time series data have been extensively exploited in the study if glacier fluctuations and glacier lake outburst [75]. Data from Landsat, ASTER, IRS, SPOT, Quickbird and IKONOS can be used for mapping and classification of glacial lakes. The topographical settings of GLOFs can be obtained from ASTER DEM, freely available ASTER GDEM and SRTM DEM [16] with high accuracy level. Huggel and others [75] have proposed an automatic methodology for mapping of Himalayan glacier lakes employing Landsat TM data. The method is known as normalized difference water index (NDWI) and uses TM1 and TM4 for distinguishing the lakes. NDWI is given as

$$NDWI = \frac{TM \text{ bands } 4 - TM \text{ bands } 1}{TM \text{ bands } 4 + TM \text{ bands } 1} \tag{8}$$

In order to calculate the volumetric changes of glaciers, especially the debris-covered type [76], stereo-capable data are useful. The Advanced Land Observing Satellite (ALOS) PRISM is a relatively new remote sensing satellite program (launched in 2006) that has stereo capability able to generate digital terrain models (DTMs) and 3D maps and that also offers high spatial resolution stereo-data (2.5 m). Several studies have investigated volumetric changes in glaciers in the Himalayas using ALOS data [77]. The estimation of area of potentially dangerous supra and proglacier lake area from remote sensing data can be used to find the glacier volume. The lake volume (V: ×106 m³) and lake area (A: km²) have the following relationship [78]:

$$V = 43.24 \times A^{1.5307} \tag{9}$$

Huggel and others [75] have also represented similar relationships from glacial lakes located in the Swiss Alps, including ice-dammed lakes. The relationship between the maximum depth of lakes (D_{max}: m) and lake areas can then be calculated as follows [78]:

$$D_{max} = 95.665 \times A^{0.489} \tag{10}$$

The depth, area and volume of glacier lakes, estimated from remote sensing technology, greatly felicitate in the assessment of GLOFs and maintain the early warning system. Figure 14 demonstrates the continuous growing of a moraine dammed lake located at the snout of Samudra Tapu glacier in Himachal Himalaya, India.

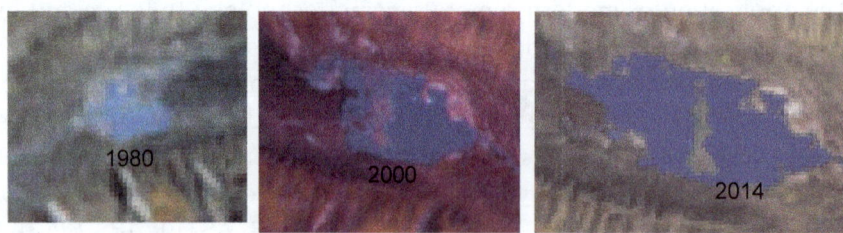

Figure 14. Growth of a moraine dammed lake in western Himalaya as shown using Landsat MSS, IRS LISS III and Landsat OLI/TIRS data of ablation season.

6.2. Snow, Ice and rock avalanches

The hazards associated with debris cover and unstable rock in the glacial environment are crucial to study as they are influenced by glacier down-wasting, glacier retreat and permafrost degradation [37] and are connected with ice avalanches and GLOFs [79]. The increasing number of ice avalanches is basically due to the changes in climatic and socioeconomic settings in the mountain region [12]. Typically an ice avalanche occurs from the surrounding steep cliffs in the glacier environment with the breaking of large mass from these cliffs and peaks. The hazards potential of ice avalanches are confined to the high mountain areas only and affect the tourists, trekkers/climbers and glaciologists. Figure 15 illustrates an avalanche prone steep slope present at the backwall of Hamtah glacier, western Himalaya. The monitoring of occurrence of ice avalanches and the settings of early warning systems for mitigation require high-quality data and tools for systematic region wide coverage. The combination of GIS tool with the remote sensing data has been found to be useful for hazard mapping in particular to debris flow and snow/ice/rock avalanches. Clague and Evans [80] have demonstrated the use of DEM for comparison of volume of ice avalanched material before and after an event. Salzmann and others [12] have shown that the glacier inventory data can be combined with the slope and aspect maps to locate the potential avalanches zones. The multitemporal data combined with DEMs can be used to identify and monitor the rock avalanches, debris flow and areas to be affected by the debris movement.

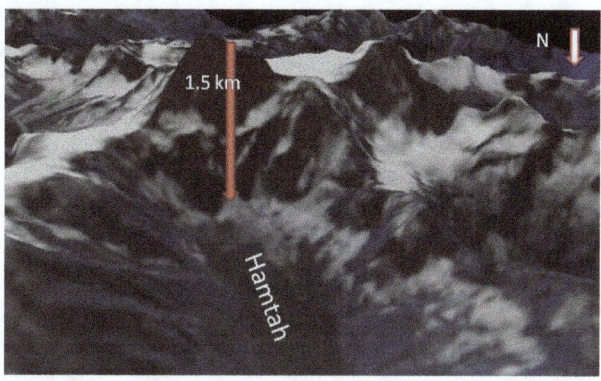

Figure 15. Landsat OLI/TIRS image dated September 28, 2014, overlaid on ASTER GDEM to show the steep back wall and surroundings of Hamtah glacier, which are susceptible to rock and ice avalanches.

6.3. Glacier surges

The glacier surges are abnormally rapid movement of large glacier parts with increased velocity due to the temporal instability of the glacier. The velocity of the glacier increases by an order of magnitude or more during the surging and the glaciers advance drastically. The glacier surges itself are not a hazard, but they induce and trigger other hazards such as ice/rock avalanches, outburst floods, blocking of river, instability of moraines and hence associated hazards. The phenomena of glacier surges are best monitored by high-frequency remote sensing data [81]. A number of glaciers in the Karakoram have been found to be showing the surging phenomena. Bhambri and others [81] have studied the surge type behavior of glaciers in Karakoram by using CORONA, Landsat TM/ETM+ data and SRTM DEM.

7. Conclusion

Glaciers are the most visible indicators of climate change, and the study of glacier parameters specifies the prevailing climate. The numbers of glaciological parameters which can be assessed and monitored by remote sensing technology are very long. The optical and radar data are equally valuable and useful for snow cover mapping, glacier area monitoring, glacier feature study, volumetric change, mass balance and velocity measurements. Optical remote sensing data is more suitable for snow cover mapping, glacier area and snout monitoring. However, glacier facies mapping, mass balance and glacier velocity can be accurately studied from radar data. DEMs are the essential requirement for studying glacier topographical and geometrical parameters. Although remote sensing methods provide an efficient tool for glacier study, the field method is the most accurate and recommended one, and remote sensing should be applied in conjunction with field work for validation.

Author details

Pratima Pandey[1*], Alagappan Ramanathan[2] and Gopalan Venkataraman[3]

*Address all correspondence to: pandeypreetu@gmail.com

1 Indian Institute of Remote Sensing, Dehradun, India

2 School of Environmental Science, Jawaharlal Nehru University, New Delhi, India

3 Centre of Studies in Resources Engineering, Indian Institute of Technology, Powai, Mumbai, India

References

[1] Ben DI and Evans DJA: *Glaciers and Glaciation.* New York: Wiley, 1998.

[2] Jansson P, Hock R, and Schneider T: The concept of glacier storage: a review. *Journal of Hydrology*, 2003, 282:116–129, doi:10.1016/S0022-1694(03)00258-0.

[3] Barnett TP, Adam JC, and Lettenmaier DP: Potential impacts of a warming climate on water availability in snow-dominated regions. *Nature*, 2005, 438(7066):303–309.

[4] Bookhagen B and Burbank DW: Toward a complete Himalayan hydrological budget: spatiotemporal distribution of snowmelt and rainfall and their impact on river discharge. *Journal of Geophysical Research*, 2010, 115(F3):F03019.

[5] Huss M, Farinotti D, Bauder A, and Funk M: Modelling runoff from highly glacierized alpine drainage basins in a changing climate. *Hydrological Processes*, 2008, 22(19): 3888–3902.

[6] Huss M: Present and future contribution of glacier storage change to runoff from macroscale drainage basins in Europe. *Water Resources Research*, 2011, 47(7):W07511.

[7] Reinwarth O and Stäblein G: Die Kryosphäre – das Eis der Erde und seine Untersuchung. *Würzburger Geographische Arbeiten*, 1972, 36:71 pp.

[8] Meier MF and Bahr DB: Counting glaciers: use of scaling methods to estimate the number and size distribution of the glaciers on the world. Edited by Hanover, NH, CRREL Spec. Rep., 1996, US Army, 89-95.

[9] Dyurgerov M and Meier MF: Glaciers and the changing earth system: a 2004 snapshot. *INSTAAR Occasional Paper*, 2005, 58:117 pp.

[10] Kääb A: Photogrammetrische Analyse zur Früherkennung gletscher- und permafrostbedinger Naturgefahren im Hochgebirge. *Mitteilungen der Versuchsanstalt für Wasserbau, Hydrologie und Glaziologie (VAW) der ETH Zürich*, 1996, 145:182 pp.

[11] Huggel C: Assessment of glacial hazards based on remote sensing and GIS modelling. PhD thesis, Department of Geography, University of Zurich, 2004, 75 pp.

[12] Salzmann ND, Kääb A, Huggel C, Allgöwer B, and Haeberli W: Assessment of the hazard potential of ice avalanches using remote sensing and GIS-modelling. *Norwegian Journal of Geography*, 2004, 58:74–84.

[13] Barrand N: Notes for Remote Sensing of the Cryosphere. http://glaciers.gi.alaska.edu/sites/default/files/Notes_remotesensing_Barrand2014.pdf

[14] Paterson WSB: *The Physics of Glaciers*, 3rd edition. Oxford: Pergamon Press, 1994.

[15] Kuhn M: The formation and dynamics of glaciers. *Remote Sensing of Glaciers*, Editors: Pellikka P and Rees G, CRC Press, Boca Raton, Fla,, 2010, pp. 21–39.

[16] Worni R, Huggel C, and Stoffel M: Glacial lakes in the Indian Himalayas — from an area-wide glacial lake inventory to on-site and modeling based risk assessment of critical glacial lakes. *Science of the Total Environment*, 2013, 468–469:S71–S84.

[17] Gareth RW and Pellikka P: Principles of remote sensing.. *Remote Sensing of Glaciers*, Editors: Pellikka P and Rees G, CRC Press, Boca Raton, Fla, 2010, pp. 1–20.

[18] Lepparanta M and Granberg HB: Physics of glacier remote sensing. *Remote Sensing of Glaciers*, Editors: Pellikka P and Rees G, CRC Press, Boca Raton, Fla,, 2010, 81–98.

[19] Rignot E, Echelmeyer K, Krabill W: Penetration depth of interferometric synthetic-aperture radar signals in snow and ice. *Geophysical Research Letters*, 2001, 28(18):3501–3504.

[20] Krieger G, Moreira A, Fiedler H, Hajnsek I, Werner M, Younis M, and Zink M: TanDEM-X: a satellite formation for high-resolution SAR interferometry. *IEEE Transactions on Geoscience and Remote Sensing*, 2007, 45(11), 3317–3341.

[21] Erasmi S, Rosenbauer R, Buchbach R, Busche T, and Rutishauser S: Evaluating the quality and accuracy of TanDEM-X digital elevation models at archaeological sites in the Cilician Plain, Turkey. *Remote Sensing*, 2014, 6:9475–9493, doi:10.3390/rs6109475.

[22] Konig M: Measuring snow and glacier ice properties from satellite. *Reviews of Geophysics*, 2001, 39 (1), 1–27.

[23] Kumar V, Venkataraman G, and Rao YS: SAR interferometry and speckle tracking approach for glacier velocity estimation using ERS-1/2 and TerraSAR-X spotlight high resolution data. Geoscience and Remote Sensing Symposium, 2009 IEEE International, IGARSS 2009 (Volume:5), Cape Town, pp. 332–335.

[24] Balmer R and Hartle P: Synthetic aperture radar interferometry. *Inverse Problems*, 1998, 14:R1–R54.

[25] Jonathan M, Infant J, Lakhankar T, Khanbilvardi R, Romanov P, Krakauer N, and Powell AL: Synergistic use of remote sensing for snow cover and snow water equivalent estimation. *British Journal of Environment & Climate Change*, 2013, 3(4):612–627.

[26] Dozier J, Painter TH, Rittger K, and Frew JE: Time–space continuity of daily maps of fractional snow cover and albedo from MODIS. *Advances in Water Resources*, 2008, 31(11):1515–1526.

[27] Dozier J: Spectral signature of alpine snow cover from the Landsat Thematic Mapper. *Remote Sensing of Environment*, 1989, 28(August 1988):9–22.

[28] Nolin AW: Recent advances in remote sensing of seasonal snow. *Journal of Glaciology*, 2010, 56(200):1141–1150.

[29] Chang ATC, Foster JL, and Hall DK: Nimbus-7 SMMR derived global snow cover parameters. *Annals of Glaciology*, 1987, 9, 39–44.

[30] Nolin AW: Recent advances in remote sensing of seasonal snow. *Journal of Glaciology*, 2010, 56(200), 1141-1150.

[31] Tedesco M and Narvekar PS: Assessment of the NASA AMSR-E SWE product. *IEEE Journal of Selected Topics in Applied Earth Observations and Remote Sensing*, 2010, 3(1): 141–159.

[32] Rott H, et al: Cold regions hydrology high resolution observatory for snow and cold land processes. *Proceedings of the IEEE*, 2010, 98(5):752–765.

[33] Raj KBG and Fleming K: Surface temperature estimation from Landsat ETM data for a part of the Baspa Basin, NW Himalaya, India. *Bulletin of Glaciological Research*, 2008, 25:19–26.

[34] Barsi JA, Barker JL, and Schott JR: An atmospheric correction parameter calculator for a single thermal band earth-sensing instrument. IGARSS 03, 21–25 July 2003, Centre de Congres Pierre Baudis, Toulouse, France, p. 3.

[35] PFEFFER WT, et al: The Randolph Glacier Inventory: a globally complete inventory of glaciers. *Journal of Glaciology*, 2014, 60(221), doi: 10.3189/2014JoG13J176.

[36] Paul F and Kääb A: Perspectives on the production of a glacier inventory from multispectral satellite data in Arctic Canada: Cumberland Peninsula, Baffin Island. *Annals of Glaciology*, 2005, 42:59–66.

[37] Kääb A: Combination of SRTM3 and repeat ASTER data for deriving alpine glacier flow velocities in the Bhutan Himalaya. *Remote Sensing of Environment*, 2005, 94(4): 463–474.

[38] Bolch, T, Menounos B, and Wheate R: Landsat-based inventory of glaciers in western Canada, 1985–2005. *Remote Sensing of Environment*, 2010, 114:127–137.

[39] Paul F, et al: The glaciers climate change initiative: methods for creating glacier area, elevation change and velocity products. *Remote Sensing of Environment*, 2013. http://dx.doi.org/10.1016/j.rse.2013.07.043

[40] Paul F and Hendriks J: Optical remote sensing of glacier extent. In:. *Remote Sensing of Glaciers*, Editors: Pellikka P and Rees G, CRC Press, Boca Raton, Fla,, 2010, pp. 137–152.

[41] Andreassen L, Paul F, Kääb A, and Hausberg J: Landsat-derived glacier inventory for Jotunheimen, Norway, and deduced glacier changes since the 1930s. *The Cryosphere*, 2008, 2(2):131–145.

[42] Rau F, Braun M, Friedrich M, Weber F, and Gobmann H: Radar glacier zones and their boundaries as indicator of glacier mass balance and climate variability. Proceedings of the EARSeL-SIG-Workshop Land Ice and Snow, Dresden, Germany, 16–17 June, 2000, pp. 317–327.

[43] Partington KC: Discrimination of glacier facies using multi-temporal SAR data. *Journal of Glaciology*, 1998, 44:42–53.

[44] Davies BJ, Carrivick J, Glasser NF, Hambre MJ, and Smellie JL: A new glacier inventory for 2009 reveals spatial and temporal variability in glacier response to atmospheric warming in the Northern Antarctic Peninsula, 1988–2009. *The Cryosphere Discussions*, 2011, 5:3541–3595.

[45] Debeer CM and Sharp MJ: Recent changes in glacier area and volume within the southern Canadian Cordillera. *Annals of Glaciology*, 2007, 46:215–221.

[46] Way RG, Bell T, and Barrand NE: An inventory and topographic analysis of glaciers in the Torngat Mountains, northern Labrador, Canada. *Journal of Glaciology*, 2014, 60(223), doi: 10.3189/2014JoG13J195.

[47] Arendt A, Echelmeyer KA, and Harrison, WD: Rapid wastage of Alaska glaciers and their contribution to rising sea level. *Science*, 2002, 297:382–386.

[48] Gardelle J, Berthier E, and Arnaud Y: Slight mass gain of Karakoram glaciers in the early twenty-first century. *Nature Geoscience*, 2012, 5(5):322–325.

[49] Cox LH and March RS: Comparison of geodetic and glaciological mass-balance techniques, Gulkana Glacier, Alaska, USA. *Journal of Glaciology*, 2004, 50(170):363–370.

[50] Soruco A, Vincent C, Francou B, Ribstei NP, Berge T, Sicart JE, Wagnon P, Arnaud Y, Favier Y, and Lejeune, Y: Mass balance of Glaciar Zongo, Bolivia, between 1956 and

2006, using glaciological, hydrological and geodetic methods. *Annals of Glaciology*, 2009, 50:1–8.

[51] Zemp M, Jansson P, Holmlund P, Gärtner-Roer I, Koblet T, Thee P, and Haeberli, W: Reanalysis of multi-temporal aerial images of Storglaciären, Sweden (1959–1999). Part 2: Comparison of glaciological and volumetric mass balances. *The Cryosphere*, 2010, 4(3):345–357.

[52] Huss M: Density assumptions for converting geodetic glacier volume change to mass change. *The Cryosphere*, 2013, 7:877–887.

[53] Kääb A, Berthier E, Nuth C, Gardelle J, and Arnaud, Y: Contrasting patterns of early twenty-first-century glacier mass change in the Himalayas. *Nature*, 2012, 488(7412): 495–498.

[54] Berthier E, Arnaud Y, Kumar R, Ahmad S, Wagnon P, and Chevallier P: Remote sensing estimates of glacier mass balances in the Himachal Pradesh (Western Himalaya, India). *Remote Sensing of Environment*, 2007, 108(3):327–338.

[55] Kulkarni AV: Mass balance of Himalayan glaciers using AAR and ELA methods. *Journal of Glaciology*, 1992, 38:101–104.

[56] Meier MF and Post AS: Recent variations in mass net budgets of glaciers in western North America. *International Association of Scientific Hydrology Publication*, 1962, 58:63–77 (Symposium at Obergurgl 1962 – Variations of Glaciers).

[57] Berthier E and Vincent C: Relative contribution of surface mass balance and ice flux changes to the accelerated thinning of the Mer de Glace (Alps) over 1979–2008. *Journal of Glaciology*, 2012, 58(209):501–512.

[58] Heid T and Kääb A: Evaluation of existing image matching methods for deriving glacier surface displacements globally from optical satellite imagery. *Remote Sensing of Environment*, 2012, 118:339–355.

[59] Berthier E, Vadon H, Baratoux D, Arnaud Y, Vincent C, Feigl KL, et al: Surface motion of mountain glaciers derived from satellite optical imagery. *Remote Sensing of Environment*, 2005, 95:14–28.

[60] Kääb A, Lefauconnier B, and Melvold K: Flow field of Kronebreen, Svalbard, using repeated Landsat 7 and ASTER data. *Annals of Glaciology*, 2006, 42(1):7–13.

[61] Luckman A, Quincey D, and Bevan S: The potential of satellite radar interferometry and feature tracking for monitoring flow rates of Himalayan glaciers. *Remote Sensing of Environment*, 2007, 111:172–181.

[62] Quincey DJ, Copland L, Mayer C, Bishop M, Luckman A, and Belò M: Ice velocity and climate variations for Baltoro Glacier, Pakistan. *Journal of Glaciology*, 2009, 55(194):1061–1071.

[63] Rankl M, Vijay S, Kienholz C, and Braun M: Glacier changes in the Karakoram region mapped by multi-mission satellite imagery. *The Cryosphere*, 2013, 7:4065–4099.

[64] Goldstein RM, Engelhart H, Kamb B, and Frolich RM: Satellite radar interferometry for monitoring ice sheet motion: application to an Antarctic ice stream. *Science*, 1993, 262(5139):1525–1539.

[65] Haeberli W and Beniston M: Climate change and its impacts on glaciers and permafrost in the Alps. *Ambio*, 1998, 27:258–265.

[66] Richardson SD and Reynolds JM: An overview of glacial hazards in the Himalayas. *Quaternary International*, 2000, 65/66:31–47.

[67] Reynolds JM: Managing the risks of glacial flooding at hydro plants. *Hydro Review Worldwide*, 1998, 6:2–6.

[68] Taylor P and Richardson SD: Glacial risks and high mining. *Mining Magazine*, 2003, April, 174–176.

[69] Kääb A, Wessels RL, Haeberli W, Huggel C, Kargel JS, and Khalsa SJS: Rapid ASTER imaging facilitates timely assessment of glacier hazards and disasters. *EOS, Transactions of the American Geophysical Union*, 2003, 84:117–121.

[70] Quincey DJ, Lucas RM, Richardson SD, Glasser NF, Hambrey MJ, and Reynolds JM: Optical remote sensing techniques in high-mountain environments: application to glacial hazards. *Progress in Physical Geography*, 2005, 29(4):475–505.

[71] Bajracharya B, Shrestha AB, and Rajbhandari L: Glacial lake outburst floods in the Sagarmatha region. *Mountain Research and Development*, 2007, 27(4):336–344.

[72] Bolch T, Buchroitner M, Peters J, Baessier M, and Bajracharya S: Identification of glacier motion and potentially dangerous glacial lakes in the Mt. Everest region/Nepal using spaceborne imagery. *Natural Hazards and Earth System Sciences*, 2008, 8:1329–1340.

[73] Wang X, Shiyin L, Wanqin G, and Junli X: Assessment and simulation of glacier lake outburst floods for Longbasaba and Pida Lakes, China. *Mountain Research and Development*, 2008, 28(3/4):310–317.

[74] Mool PK: Glacier lake outburst floods in Nepal. *Journal of Nepal Geological Society*, 1995, 11:273–280.

[75] Huggel C, Kääb A, Haeberli W, Teysseire P, and Paul F: Remote sensing based assessment of hazards from glacier lake outbursts: a case study in the Swiss Alps. *Canadian Geotechnical Journal*, 2002, 39(2):316–330.

[76] Sawagaki T, Damodar L, Alton CB, and Teiji W: Changes in surface morphology and glacial lake development of Chamlang South Glacier in the eastern Nepal Himalaya since 1964. *Global Environmental Research*, 2012, 16(2012):83–94.

[77] Lamsal D, Sawagaki T, and Watanabe: Digital terrain modelling using Corona and ALOS PRISM data to investigate the distal part of Imja Glacier, Khumbu Himal, Nepal. *Journal of Mountain Science*, 2011, 8:390–402.

[78] Sakai A: Glacial lakes in the Himalayas: a review on formation and expansion processes. *Global Environmental Research*, 2012, 16(2012):23–30.

[79] Huggel C, Haeberli W, Kääb A, Bieri D, and Richardson S: An assessment procedure for glacial hazards in the Swiss Alps. *Canadian Geotechnical Journal*, 2004, 41:1068–1083.

[80] Clague JJ and Evans SG: A review of catastrophic drainage of moraine-dammed lakes in British Columbia. *Quaternary Science Reviews*, 2000, 19:1763–1783.

[81] Bhambri R, Bolch T, Kawishwar P, Dobhal DP, Srivastava D, and Pratap B: Heterogeneity in glacier response in the upper Shyok valley, northeast Karakoram. The Cryosphere, 2013, 7:1385–1398, doi:10.5194/tc-7-1385-2013.

2

Optical Satellite Remote Sensing of the Coastal Zone Environment

Ana C. Teodoro

Additional information is available at the end of the chapter

Abstract

Optical remote-sensing data are a powerful source of information for monitoring the coastal environment. Due to the high complexity of coastal environments, where different natural and anthropogenic phenomenon interact, the selection of the most appropriate sensor(s) is related to the applications required, and the different types of resolutions available (spatial, spectral, radiometric, and temporal) need to be considered. The development of specific techniques and tools based on the processing of optical satellite images makes possible the production of information useful for coastal environment management, without any destructive impacts. This chapter will highlight different subjects related to coastal environments: shoreline change detection, ocean color, water quality, river plumes, coral reef, alga bloom, bathymetry, wetland mapping, and coastal hazards/vulnerability. The main objective of this chapter is not an exhaustive description of the image processing methods/algorithms employed in coastal environmental studies, but focus in the range of applications available. Several limitations were identified. The major challenge still is to have remote-sensing techniques adopted as a routine tool in assessment of change in the coastal zone. Continuing research is required into the techniques employed for assessing change in the coastal environment.

Keywords: Shoreline Change Detection, Ocean Color, Optical Water Quality, River Plumes, Coral Reef, Alga Bloom, Bathymetry, Wetland Mapping, Hazards, Vulnerability

1. Introduction

One of the most useful reviews of remote sensing of the coastal zone was the work published by Cracknell [1], where a review of the current state of the use of remote sensing in estuaries and coastal waters at the end of the 20th century was performed. He identified that period (end of the 20th century) as a stage of potential great changes and advances in the use of remote sensing. Since then, the advances in the use of remote sensing for coastal areas have been huge. These advances are related to the availability of new sensors, more adequate for the study of this area, and also the improvements in the classification algorithms. Several useful reviews related to the value of remote sensing in the coastal zone environment have been published since then [2, 3]. Malthus and Mumby [2] update the information given by Cracknell [1], and highlight a number of priority areas. Advances were identified in the benefit of high spatial and spectral resolution data and complementary remote-sensing techniques. Further benefits are identified in rapid and more frequent data acquisition, faster and more automated processing and a greater sampling intensity over conventional field-based techniques. All these aspects were fully confirmed. Issues associated with adoption of remotely sensed data for coastal management were also discussed. This issue still is a topic of extreme importance. Although remotely sensed data are currently used for decision-making, their use is not yet an integrated tool for coastal management. Several research priorities were identified in the work of Malthus and Mumby [2]. Areas of value that continue to remain poorly investigated include the improvements to be gained from synergistic use of multiwavelength remote-sensing approaches, change detection techniques, and multitemporal comparisons and knowledge-based approaches to improve classification [2]. The lack of accuracy remains a challenge task. Therefore, the major challenge is to implement the remote-sensing techniques as a routine tool in assessment of coastal zone changes. Unfortunately, this challenge is still unfulfilled, as will be described in this chapter. More recently, Klemas [3] published an overview of remote sensing of emergent and submerged wetlands. Kelmas [3] discusses the impact of climate change on coastal wetland (sea-level rise, increase of temperature, and changes in precipitation), and the impacts due to anthropogenic activities. He has enumerated the recent advances in sensor design (high-resolution multispectral and hyperspectral imagers, light detection and ranging (LiDAR), and radar systems), and image processing techniques that making remote-sensing systems more practical and attractive for monitoring coastal ecosystems. The lack of accurate near-shore bathymetric data was identified as a key limitation in the application of geospatial data to coastal environments. He concludes that when remote-sensing systems are used wisely, including complementary combinations of different satellite and airborne sensors, they can provide data that enhance the research and management of coastal ecosystems. According to Klemas [3], the future research priorities should include better understanding and description of the radiative properties of coastal environments. Additional knowledge is required about the spatial and temporal variations of water column optical properties and its constituents. Best approaches for processing hyperspectral data need to be further investigated [3].

The main objectives of this chapter are (i) to provide an overview of the optical satellite remote sensing of the coastal zone environment and (ii) to highlight a number of application fields related to coastal areas where optical remote sensing plays an important role.

2. Optical remote sensing for coastal areas: Principles

Optical imaging sensors are a crucial technology in the field of coastal remote sensing. The main function of electro-optical imaging sensors is to collect incident electromagnetic (EM) radiation and convert it to a stored representation useful for remote-sensing analysis. These sensors operate in the optical region of the EM spectrum defined as radiation with wavelengths between 400 and 15000 nm. This range includes the visible (400–700 nm), the near infrared (NIR, 700–1100 nm), the short infrared (SWIR, 1100–2500 nm), the midwave infrared (MWIR, 2500–7500 nm) and the long-wave infrared (LWIR, 7500–15000 nm) spectral regions [4]. Optical remote sensing involves acquisition and analysis of optical data-EM radiation captured by the sensing modality after reflecting off an area of interest on ground/water. Different materials/ water constituents reflect and absorb differently at different wavelengths. Thus, the targets/ elements can be differentiated by their spectral reflectance signatures in the remotely sensed images. The optically active water constituents, including phytoplankton (chlorophyll a – Chla), detritus and minerals, Colored Dissolved Organic Matter (CDOM – also called gelbstoff or yellow substances), and water itself, all have an impact on the optical signature of water in the visible wavelengths. In the visible spectral range of solar radiation, light can penetrate in water bodies and its color can change due to scattering and absorption processes in the water body or at its bottom. This makes it possible to derive from optical remote-sensing data information about the characteristics of the water body and the type/concentration of its components. The water curve (spectral signature) is characterized by a high absorption at NIR wavelengths range and beyond. Because of this absorption property, water bodies as well as features containing water can easily be detected, located, and delineated with remote-sensing data. Turbid water has a higher reflectance in the visible region than clear water. This is also true for waters containing high Chla concentrations. Coastal waters are optically complex and the signal that a remote detector collects is a mixed signal including various water optically active constituents from different sources. Complex interaction among phytoplankton (Chla), Total Suspended Mater (TSM), and CDOM results in poor predictive ability in retrieval of various water quality proprieties in coastal waters.

Optical remote-sensing systems are classified into different types, depending on the number of spectral bands used in the imaging process: 1) Panchromatic imaging system: the sensor is a single-channel detector sensitive to radiation within a broad wavelength range. If the wavelength range coincides with the visible range, then the resulting image resembles a "black-and-white" image. 2) Multispectral imaging system: the sensor is a multichannel detector with a few spectral bands. Each channel is sensitive to radiation within a narrow wavelength band. The resulting image is a multilayer image which contains both the brightness and spectral information of the targets. 3) Hyperspectral Imaging Systems: the sensor acquires images in several (typically hundred or more) contiguous spectral bands. The precise spectral information contained in a hyperspectral image enables better characterization and identification of targets. Hyperspectral images have a great potential in applications regarding coastal management.

3. Sensors

In coastal and inland waters, optically active constituents often vary independently requiring improved spectral and radiometric resolutions, while physical drivers such as tides and geographic boundaries set up different spatial and temporal scales compared to the open ocean [5]. Due to the large number of sensors available, with distinct characteristics, it is a challenge to choose the most appropriate satellite images for monitoring coastal environments. The selection of the sensor is related to the applications required and the different types of resolution (spatial, spectral, radiometric, and temporal) should be considered. Another aspect that could interfere with the selection of the sensor is the data availability. Some images are really expensive and some data can be freely downloaded or granted by national/international organizations for research purposes. A list of the most relevant optical sensors used in the last decade to the assessment of coastal zone environment is shown in Table 1. A number of sensors have been launched since the Coastal Zone Color Scanner (CZCS) in 1978, including the Sea-viewing Wide Field-of-viewSensor (SeaWiFS), the MODerate resolution Imaging Spectrora-diometer (MODIS), and the MEdium Resolution Imaging Spectrometer (MERIS). These instruments are equipped with sensors optimized for measuring water-leaving radiance or reflectance over most of the world's oceans, but not over many inland or coastal waters. Recently, significant advances have been made in studying coastal and inland waters using global sensors such as MODIS medium resolution data and MERIS full resolution (FR) data [6-8]. The primary mission of MERIS was the measurement of sea color in the oceans and in coastal areas. The applicability of MERIS data to coastal studies is extensive. Unfortunately, the MERIS instrument is no longer available (since May 2012).

Traditionally, the Landsat (TM and ETM+), the French Système Pour l'Observation de la Terre (SPOT), and Terra/ASTER have been reliable data sources for large coastal watersheds' land-cover [9, 10], water turbidity quantification [11], suspended sediments' concentration estima-tion [12-15], vegetation cover [16], among others. However, the 30 m, 20 m, and 15 m, respectively, spatial resolutions in the visible and Near Infra-Red (NIR) bands were initially designed for land-cover studies. The availability of high spatial and spectral resolution satellite data has significantly improved the capacity for mapping coastal ecosystems. High-resolution imagery obtained from satellites, such as IKONOS-2, Quick Bird-2, GeoEye-1, and Orbview-3 can be used for different purposes regarding coastal applications. WorldView-2 has a spatial resolution of 2 m for 8 multispectral (MS) bands (4 standard colors: red, blue, green, NIR, and 4 new colors: red edge, coastal, yellow, NIR2, and 0.5 m spatial resolution for the panchromatic (PAN) band (450–800 nm). The Pleiades 1A/1B satellites were designed with urgent tasking option, and images can be requested less than six hours before they are acquired. This functionality will prove invaluable in situations where the expedited collection of new image data is crucial, such as coastal crisis monitoring. This sensor is comparable to the other high-resolution sensors (e.g., GeoEye-1, Orbview-3). The Hyperion provides a high-resolution hyperspectral imager capable of resolving 220 spectral bands with a 30 m resolution. Through these spectral bands, complex coastal ecosystems can be imaged and accurately classified.

Sensor	Spectral Range (nm)	No. Bands	Spatial Resolution	Temporal Resolution	Swath width
Landsat TM	450–900	4 VNIR	30 m	16 days	185 km
	1550–2350	2 SWIR	30 m		
	10410–12500	1 TIR	120 m		
Landsat ETM+	450–900	4 VNIR	30 m	16 days	183 km
	1550–2350	2 SWIR	30 m		
	10410–12500	1 TIR	60 m		
	520–900	1 PAN	15 m		
SPOT 4-5 HRVIR	500–890	3 VNIR	20 m	26 days	60 km
	1580–1750	1 SWIR	20 m		
	610–680	1 PAN	10 m		
SPOT 5 HRS	500–890	3 VNIR	10 m	26 days	60 km
	1580–1750	1 SWIR	20 m		
	510–730	1 PAN	5 m		
ASTER	520–860	3 VNIR	15 m	16 days	60 km
	1600–2430	6 SWIR	30 m		
	8125–11650	5 TIR	90 m		
MODIS	620–14385	16 VNIR	250 m–1 km	1 day	2330 km
		4 SWIR			
		16 TIR			
SeaWIFS	402–885	8 VNIR	1.1 km	1 day	2800 km
MERIS	290–1040	15 VNIR	300 m	<3 days	1150 km
Hyperion EO-1	400–2500	220	30 m	16 days	8 km
IKONOS-2	455–850	4 VNIR	4 m	1–3 days	11 km
	760–850	1 PAN	1 m		
Quick Bird	430–918	4 VNIR	2.44 m	<3 days	16.5 km
	405–1053	1 PAN	0.61 m		
Orbview-3	450–900	4 VNIR	4m	<3 days	8 km
	450–900	1 PAN	1m		
GeoEye-1	450–920	4 VNIR	1.65 m	2.1–8.3 days	15.2 km
	450–800	1 PAN	0.41 m		
WorldView-2	400–1040	8 VNIR	1.85 m	1.1–2.7 days	16.4 km
	450–800	1 PAN	0.46 m		
Pleiades 1A/1B	430–950	4 VNIR	2.0 m	1 day	20 km
	480–830	1 PAN	0.5 m		
Sentinel-2	420–2370	VNIR-SWIR	10,20, 60 m	<3 days	290 km

Table 1. Characteristics of some optical systems used in coastal zones applications

Shortly, the assessment to the Sentinel-2 data will improve coastal environment monitoring programs. The Sentinel-2 was launched in June 2015 within COPERNICUS programme of the European Space Agency (ESA). The design of the Sentinel-2 mission aims at an operational multispectral Earth-observation system that complements the Landsat and SPOT and improves data availability for users. More information about Sentinel-2 can be found in Drusch et al. [17].

The development of specific techniques based on the processing of optical satellite data makes possible the production of information really useful for coastal environments, without any destructive impacts. Different image processing techniques have been applied to the satellite images in order to study the coastal environment. These techniques differ depending on the subject of study. Most of the techniques widely used in land and ocean studies are also applied in coastal research. Some techniques have also been intentionally developed to study specific aspects of this area. The topic of this chapter is not an exhaustive description of the image processing methods/algorithms employed in coastal environmental studies, but focus in the range of applications available. In this chapter will be gathered the most cited/important applications of optical remote sensing regarding the coastal zone environment of the last decade.

4. Applications

In this section, several application fields related to coastal environments, where optical remote sensing plays an important role, are addressed.

4.1. Shoreline change detection

Shorelines are inherently dynamic features that mark the transition between land and sea and are vulnerable to waves, winds, nearshore currents, and anthropogenic actions [18]. It is estimated that there are around 350 000 km of shoreline in the world and more than 60% of the world's population lives within 100 km of the coastal/sea. Therefore, monitoring and managing shorelines evolution are of considerable social, cultural, and economic importance. Furthermore, shoreline erosion and coastal flooding were highlighted among the gravest effects of climate change [19]. Several studies have investigated the potential of optical satellite images to study shoreline change. An idealized definition of shoreline is that it coincides with the physical interface of land and water [20]. Because of the dynamic nature of the idealized shoreline boundary, the use of shoreline indicators has been adopted for coastal studies. A shoreline indicator is a feature that is used as a proxy to represent the "true" shoreline position. Boak and Turner [21] reviewed shoreline definition and detection techniques, and carried out a comprehensive literature study. They categorized shoreline indicators in three groups: (i) visible discernible features; (ii) tidal datum-based indicators; and (iii) indicators based on the processing technique to extract the shoreline. One of the most common technique for shoreline detection was (and still is in some cases) visual interpretation. However, this approach is highly subjective and is not possible to access to any accuracy indicator. The alternative employs digital image processing techniques, as supervised and unsupervised classification algo-rithms. Gen [22] presents a paper that reviews the status of the use of remote sensing for the detection, extraction, and monitoring of coastlines. The review takes the US system as an example. However, the issues researched can be applied to any other part of the world. He concludes that visual interpretation of airborne remote-sensing data is still widely and popularly used for coastal delineation. However, a variety of remote-sensing data and techniques are available to detect, extract, and monitor the coastline.

Guariglia et al. [23] used a multisource approach to coastline mapping, in Basilicata region (Italy). They stated that satellite images are affected by tidal variations depending on their spatial resolution and concluded that the coastline can be extracted from Landsat TM images, without the interference of the tidal factor. Instead, tidal effects must be considered when the coastline is identified from images having higher spatial resolution that are comparable to the errors induced by tide.

Ekercin [24] present a work on the coastline movements at the northeast coasts of the Aegean Sea (Turkey). In this study, the coastline changes were examined using data from Landsat MSS, TM, and ETM collected between 1975 and 2001. In the image processing step, an unsupervised image classification algorithm (ISODATA) was employed and temporal image ratioing techniques were used to carry out coastline change assessment. Significant coastline move-ments were identified.

Maiti and Bhattacharya [25] used multidate satellite images from Landsat MSS, TM, ETM+, and ASTER to demarcate shoreline positions, from which shoreline change rates have been estimated using linear regression, along the coast of Bay of Bengal (India), between 1973 and 2003. The shorelines have been identified through the NIR bands, and included gray level thresholding and segmentation by edge enhancement technique. The result shows that 39% of transects have uncertainties in shoreline change rate estimations. On the other hand, 69% of transects exhibit lower Root Mean Square Error (RMSE) values for the short-term period, indicating better agreement between the estimated and satellite-based shoreline positions.

Kuleli et al. [26] presented a research focused on the shoreline change rate analysis by automatic image analysis techniques through histogram-based segmentation of land and water based on automatic thresholding algorithm, using multitemporal Landsat images (MSS, TM and ETM+) between 1972 and 2009 along the coastal Ramsar wetlands of Turkey. Accretion or erosion processes were observed on multitemporal satellite images along the areas of interest.

Kumar et al. [27] applied and developed the method established by Maiti and Bhattacharya [25] for calculating the rates of shoreline change, shoreline positions, and morphology of spits along the Karnataka coast, western India, for the period from 1910 to 2005 using multidated satellite images and topographic maps. Satellite images (IRS 1C, LISS-III) of IR band were employed. Binary images are used as input layers in unsupervised classification module to a complete separation between land and water classes, and to remove effect of suspended materials, if any. Significant changes in morphology of spits have been recorded.

Wang et al. [28] presented a class association rule algorithm on the basis of the Apriori algorithm. To test the feasibility of the method, Landsat ETM+ image scene of Jiaozhou Bay near Qingdao city (China) was used to interpret the coastline. First, the association rules of the sea–land separation of the study area were discovered from learning samples by using the class association rule algorithm. Second, the sea and the land of the image were separated with the mined rules. Third, the coastline was interpreted from the separation result. This approach includes not only spectral attributes but also the texture attributes (entropy) and the statistical analysis variables (mean and variance).

Regarding sand spits' behavior, Teodoro and Gonçalves [29] present different approaches in order to extract sand spits from IKONOS-2 data (Figure 1). A semiautomatic approach is proposed in this work, which is based on global thresholding through the Otsu's method, further refined through detected edges (GThE). The performance of GThE is compared with traditional pixel-based and object-based classification algorithms. The dataset is composed by six IKONOS-2 images, acquired between 2001 and 2007, covering a sand spit located in Portugal. The performance of the different methods used in the estimation of the sand spit area was evaluated through two sets of reference values of the sand spit area. The proposed GThE method presented better results than the other traditional methods, with a clear advantage of a considerable faster performance, beyond requiring a minimum operator intervention.

A high-precision geometric method for automated shoreline detection in the Spanish Mediterranean coast, from 45 Landsat TM and ETM+ imagery was presented by Pardo-Pascual et al. [30]. The methodology is based on an algorithm for subpixel shoreline extraction. The algorithm is based on the assumption that the separation between water and land will occur where the infrared intensity gradient around the pixel-level shoreline is maximum. The results confirm that the use of Landsat imagery for detection of instantaneous coastlines yields accuracy comparable to high-resolution techniques.

More recently, García-Rubio et al. [31] developed a method to identify the shoreline from satellite optical images (SPOT), applying an unsupervised classification (ISODATA), using the NIR spectral band to separate the sea and the land in Progreso (Yucatán, México). The shoreline was validated using quasi-simultaneous in situ shoreline measurements, both adjusted to equal water levels. The validation of shoreline obtained by satellite data revealed that the shoreline is located consistently seaward of the in situ shoreline. The success of this method suggests that it should be applicable to other locations, after adapting the confidence bounds to the beach conditions.

In conclusion, several techniques for coastline extraction and change detection from optical satellite imagery have been developed in the recent years. Manual identification, image enhancement, density slice using single or multiple bands, and image classification (supervised and unsupervised) are still the most common techniques employed. In addition, several image processing methods related to segmentation algorithms and statistics approaches have also been used. The data more used still are the traditional Landsat and SPOT images, but some works had also used high spatial resolution data (e.g., IKONOS 2), regarding the availability of an NIR band. In the future, should be considered the recent availability of the new sensors in conjunction with classification/segmentation algorithms more efficient. The

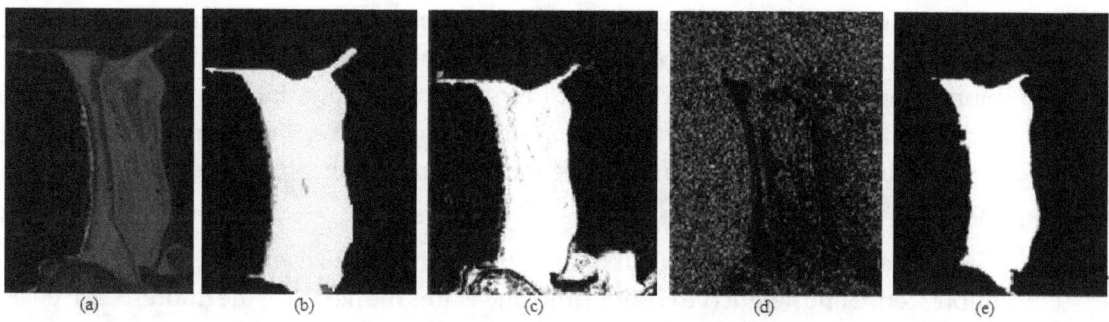

Figure 1. (a) Panchromatic band of the IKONOS-2 image from Jun. 2005; (b) the sand spit extraction with object-based approach; (c) global thresholding of the image in Fig. 1(a) through the Otsu's method; (d) edges of the image in Fig. 1(a) obtained through the Canny edge detector; (e) final extraction of the sand spit in Fig. 1(a), through the refinement of the global thresholding in (c) through the edges represented in (d) (adapted from [29])

accurate extraction of the shoreline is one of the most important parameter to estimate the erosion rates.

4.2. Coastal color

Remote sensing of ocean color has an important role to play as a cost-effective tool for global and frequent observations that can be interpreted in terms of surface concentrations of Chla, TSM, or CDOM. However, this global capability is to some extent questioned by the uneven distribution of field data that are at the basis of empirical algorithms, or are used for the definition of parameters in semianalytical bio-optical algorithms, and frequently these algorithms are not calibrated for coastal waters. The dominant optically active constituent in the open sea (case-1 waters) is the Chla, whereas in coastal waters (case-2 waters), TSM and CDOM often dominate the spectral signal of Chla [32].

Chlorophyll-a (Chla)

Chla is certainly the most commonly derived parameter in water quality mainly because of its use in determining the trophic status of waters. The Chla estimation allows forecasting of the phytoplankton concentration and is therefore an important component in the derivation of secondary products such as primary production. Several techniques/algorithms have been applied in order to estimate the Chla concentration [33]:

i.   In high-biomass waters the 700/670 nm ratio reflectance has been widely used. The explanation for the strength of the correlation of Chla with the 700/670 nm is based on the interaction between backscattering from phytoplankton and the strong absorption of water, which both increase toward the IR. The offset to scattering due to absorption by water near 700 nm causes a sharp peak in highly scattering waters. The height and position of this peak is known to be well-correlated with Chla, with the peak shifting toward greater wavelengths (apx. 715 nm) as Chla increases. In contrast, the reflectance near 670 nm is uncorrelated, with Chla being almost constant owing to the Chla absorption maximum, which offsets backscattering. The position-ing of the MERIS bands at 665 and 709 nm makes MERIS ideally suited for predicting

Chla using this ratio, and many studies have recently been carried out [34, 35]. Three-band algorithm has also been used to estimates of Chla in turbid and very high biomass hypertrophic waters [36]. A four-band algorithm, including an additional band near 700 nm, was found to be an improvement over the three-band model in highly turbid lake water through better accounting for absorption by water and nonnegligible scattering by TSM in the NIR band [37].

ii.   The fluorescence maximum near 685 nm has been used to estimate Chla [38, 39]. The fluorescence line height (FLH) algorithm measures the height of the fluorescence peak at 685 nm from a linear baseline drawn between two points on either side of the peak [40]. It is important to consider that the FLH algorithm is only suitable for Chla concentrations generally not exceeding 30 mg m^{-3} as the backscattering peak near 700 nm overwhelms the fluorescence peak in high-biomass water.

iii.   Sensors such as Landsat [41], SPOT [42], and IKONOS [43] are also frequently used to estimate Chla. However, the lack of narrow bands and low Signal–Noise Ratio (SNR) make very difficult the use of the algorithms already described. Therefore, simple linear regressions of single bands or band ratios are used and with less-significant correlations. An alternative could be the use of advanced algorithms, such as Artificial Neural Networks (ANN) and genetic algorithms [44], multivariate regression analysis [45], or spectral decomposition algorithm [46]. The use of these and other complex algorithms generally leads to improved significance of correlations.

Total Suspended Mater (TSM)

TSM is the total mass of suspended particles as measured per volume of water including inorganic (minerals) and organic (detritus and phytoplankton) components. The study of TSM concentration has a huge ecological importance, because the suspended matter is the main carrier of various inorganic and organic substances and becomes the main substrata for biochemical processes [47]. The TSM concentration affects ocean/coastal productivity, water quality, navigation, and coastal defense. The TSM concentration and distribution in the coastal zone varies with several hydrodynamic factors, such as tidal condition, currents' direction and velocity, river discharges, and wind stress [12]. The discrimination of TSM from water reflectance is based on the relationship between the scattering and absorption properties of water and its constituents. In the visible and NIR region, most of the scattering is caused by suspended sediments, and the absorption is controlled by Chla and CDOM. Therefore, the visible and NIR regions are the most adequate to estimate the TSM concentration. These absorptive in-water components decrease the reflectance in a substantial way. However, these absorptive effects occur generally for wavelengths less than 500 nm [32]. Several works have demonstrated that optical remotely sensed data can be used to retrieve TSM concentration from turbid coastal waters [14]. Many TSM models based on empirical methods have been used in operational satellite systems. These models were developed on the basis of statistical relationships between TSM concentrations and single-band or multiband reflectance [12, 13]. Although empirical models may be effectively applied to satellite images concurrent with the calibration dataset, their accuracy may be reduced outside the conditions of the calibration dataset because of the empirical basis [48]. Therefore, semianalytical models that combine physical methods with statistical methods were proposed for several authors in order to retrieve the TSM concentration [49]. Teodoro et al. [12] present different methodologies to estimate the TSM concentration in a particular area of the Portuguese coast, from remotely sensed multispectral data (ASTER, SPOT HRVIR, and Landsat TM), based on single-band models, multiple regression, and ANNs. The analysis of the RMSE achieved by both the linear and nonlinear models supports the hypothesis that the relationship between the seawater reflectance and TSM concentration is clearly nonlinear. The ANNs have been shown to be useful in estimating the TSM concentration from reflectance of visible and NIR bands of ASTER, HRVIR (Figure 2), and TM sensors, with better results for ASTER and HRVIR sensors.

Colored Dissolved Organic Matter (CDOM)

CDOM, also called gelbstoff or yellow substances, is primarily composed of humic acids produced from the decomposition of plant litter and organically rich soils within coastal watersheds and upland areas is a significant contributor to water color, because humic substances absorb strongly in the blue region of the spectrum, turning the water brown. The absorption by CDOM (aCDOM – usually referenced at 440 nm) takes the form of an exponential function decreasing toward longer wavelengths so that its effects are usually negligible at wavelengths higher than 550 nm. CDOM concentrations increase in coastal waters due to the in situ creation of fulvic acids produced from the seaweed decomposition as a by-product of primary production estimulated by nutrients and the anthropogenic input of industrial or domestic effluents from populated areas. In the coastal environment, the optical properties of CDOM change owing to seawater mixing and photodegradation. Absorption by CDOM is one of the primary additive absorption Inherent Optical Properties (IOPs), along with phytoplankton and water, and is of great interest from a bio-optical perspective. Algorithms using ratios of reflectance in visible range have been found to be well-correlated with aCDOM [50]. In waters with low TSM concentration, Bowers et al. [51] showed theoretically, while making some assumptions about particulate absorption, that there is a linear relationship between aCDOM and the ratio of reflectance in the red and blue bands. Doxaran et al. [52] used a 400/600 nm ratio, whereas D'Sa and Miller [53] used the SeaWiFS band configurations 412/510, 443/510, and 510/555 nm, all of which gave good results, although this may reflect the existence of strong covariance between Chla and CDOM. Comparable red/blue ratios produced with the MERIS data also give similarly strong correlations [54]. The low radiometric resolution of some sensors (TM, IKONOS) makes CDOM estimations infeasible [55]. More recently, Loisel et al. [56] proposed a new method to assess aCDOM, based on the theoretical link between the vertical attenuation coefficient and the absorption coefficient. This method, confirmed from radiative transfer simulations and in situ measurements, and tested on an independent in situ data set allows aCDOM to be assessed with higher accuracy.

The optically active water constituents, including Chla, TSM, CDOM, and water itself, all have an impact on the optical signature of water in the visible wavelengths. The water-leaving radiance is modified through the backscattering and absorption of light by these constituents (IOPs). Absorption by Chla, CDOM and detritus, and water itself, are well-defined in the literature and can be used to explain the causal relationships between the observed remote-sensing reflectance and the biogeophysical parameters of interest. The backscattering coefficients for water, minerals, Chla, and TSM can be used in the same way. Strong absorption by water at wavelengths >750nm effectively masks out the signals from other constituents except in highly turbid water where scattering by minerals overwhelms absorption by water. Therefore, wavelengths between 400 and 750 nm generally contain the the most important information on the water constituents, which is detectable by remote-sensing instruments, with the exception of highly turbid water where the signal in the NIR is also useful [33]. Matthews [33] present a review of the empirical procedures of remote sensing in inland and near-coastal transitional waters. A review of empirical algorithms for quantitatively estimating a variety of parameters, including Chla, TSM, turbidity, and aCDOM, were proposed. The theoretical basis of the empirical algorithms was given using fundamental bio-optical theory of the IOPs. More recently, Mouw et al. [7] presented a review that describes the current and desired state of the aquatic satellite remote sensing, namely, mission capability, in situ observations, algorithm development, and operational capacity. They concluded that significant advances have been made in supporting in situ observations, algorithm development, and operational capacity and user engagement, but challenges still exist.

One of the major challenges in coastal and inland waters is the high turbidity and strong absorption. As absorption increases, the effect of self-shading of upwelling radiance increases. For IOPs, the available scattering sensors have the capability to effectively resolve backscattering at very high levels, but standard gain settings for these sensors are typically set to saturate at levels an order of magnitude lower to maximize resolution in the dynamic ranges observed in the ocean.

4.2.1. Optical Water Quality (OWQ)

Optical water quality (OWQ) has been defined by Kirk [57] as "the extent to which the suitability of water for its functional role in the biosphere or the human environment is determined by its optical properties." There are four main natural constituents, broadly classified, that attenuate light besides water itself: CDOM, TSM, nonalgal particulate organic matter (POM), and phytoplankton. Assessing OWQ involves quantifying the behavior of light in waters as affected by these light-attenuating constituents. Several publications have described the application of optical remote-sensing systems to measure water-quality conditions in lakes [45], river systems [58], and coastal zones [59, 60]. The interpretation of optical remote-sensing data of estuaries and tidal flat areas is hampered by optical complexity and often extreme turbidity. Extremely high concentrations of TSM, Chla and CDOM, local differences, seasonal and tidal variations, and resuspension are important factors influencing the optical properties in such areas [61].

There are mainly two approaches for deriving water-quality products from remotely sensed data: the model-based and the empirical approach. The model-based (or analytical) approach seeks to model the remote-sensing reflectance in terms of the water IOPs through radiative transfer modeling [62]. The remote-sensing reflectance from the water IOPs is obtained through a bio-optical model and an approximation of the radiative transfer equation [63] or through direct solution of the Radiative Transfer Equation (RTE). The reflectance at the top of the atmosphere can then be modeled using radiative transfer calculations for the atmosphere through codes such as 6S [64]. The main concerns with these kinds of algorithms are their sensitivity to errors from atmospheric correction procedures and the existence of nonunique or ambiguous solutions arising from the additive nature of the IOPs and the consequences of using a ratio in the reflectance approximation [65]. The analytical approach is complex and requires measurements of local/regional IOPs to develop a robust forward model. Empirical algorithms are relatively simple to derive and use: simultaneously acquired experimental sets of limnological, atmospheric, and remotely sensed data are used to normally derive site-and-time specific algorithms for a certain parameter using statistical regression techniques. These algorithms generally produce robust results for the areas and data sets from which they are derived. There are many varieties of algorithms that use either single bands, band ratios, band arithmetic, or multiple bands as independent variables in linear, multiple linear, or nonlinear regression analyses [33]. The empirical approach is computationally simpler, and it is employed in the majority of studies in inland waters.

Mélin and Vantrepotte [66] presented a study about the satellite data (SeaWiFS) available for coastal/shelf waters and marginal seas to derive a set of optical water types encompassing the full extent of the optical variability found in these regions. The spatial and temporal sampling considered is well-adapted to capture the optical variability found in coastal waters, whereas a higher level of averaging would tend to smooth out peculiar spectral characteristics. The focus of this work was all the coastal regions and marginal seas of the world. The classification allows the quantification of the optical similarity between regions. The set of 16 classes used in this work covers very turbid waters founded close to river outflow regions to oligotrophic waters. The general variability in optical types at any location has been addressed by quantifying the number of classes selected as dominant during the period and an index of optical diversity that has been linked to indices of marine biodiversity.

The works referred forecast an increasingly important role for OWQ studies driven by increased awareness of the need to protect ecosystems, manage water resources, and advance remote-sensing capabilities.

4.2.2. River plumes

River discharge into the coastal waters represents a major link between terrestrial and marine systems. River plumes are a mixture of freshwater and river sediment load, with some dilution caused by currents, and are affected by many factors such as river discharge, coastal wind fields, water stratification, surface layer mixing, tides and current, etc. It is known that plume waters near river mouths can contain high concentrations of nutrients and are excessively turbid [68].

CDOM is often used as an effective tracer for evaluating relative levels and the spatial distribution of dissolved organic carbon in aquatic environments. In addition, both Chla concentrations and turbidity are typically much higher in river systems, when compared to open sea environments. The river plumes are also distinguished from surround marine waters by their high concentration TSM, which changes the color of the ocean surface. Optical satellite images have been widely employed to study the spatio temporal variations of major river plumes around the world. The river plume observations/quantification included data from

AVHRR [69], SeaWiFS [70], MODIS [71], MERIS [72], Landsat TM/ETM [74], or combining data from different sensors [73].

Zhu and Yu [75] present a study that aimed to evaluate the effectiveness of an inversion algorithm for the extraction of riverine and estuarine CDOM properties at global scales through EO-1 Hyperion images applied to ten major rivers from five continents. The river plumes are distinguished from surrounding marine waters by their high concentration of TSM which changes the color of the ocean surface. Since the TSM concentration can be associated with nutrients, pollutants, and other materials, it is of crucial importance to remotely survey their dispersal in order to assess the coastal environmental quality of the regions surrounding river mouths.

Lihan et al. [76] present a study to identify the Tokachi River plume by satellite images (SeaWiFS) and determine its relationship with river discharge and clarify its temporal and spatial dynamics. A supervised (Maximum Likelihood – ML) classifier was used to identify the plume and empirical orthogonal functions were applied to determine the spatial and temporal variability of the plume during 1998–2002.

Gonçalves et al. [4] proposed an automatic procedure for the identification of the Douro river plume (Portugal) based on the thresholding of the 71 of MERIS FR scenes (level 2 data – TSM band) through an automatic search for the optimum value of the threshold parameter. A fully automatic method was considered and a comprehensive characterization of the river plume was performed through a set of attributes, which take into account not only the shape of the river plume, but also the TSM concentration values. Regarding the characterization in terms of shape, the following attributes were considered: size of the plume; corresponding to the number of pixels; perimeter; the major and the minor axis length of the ellipse adjusted to the river plume; and the orientation (Fig. 3).

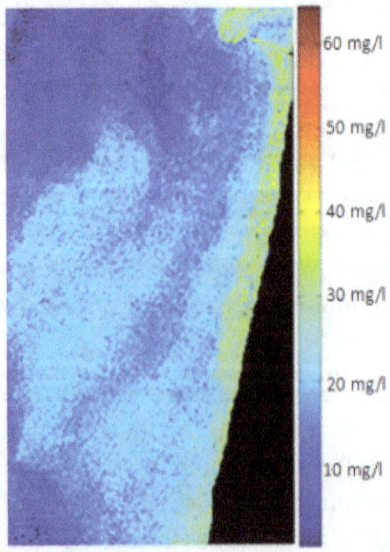

Figure 2. TSM concentration estimated from the HRVIR image through ANN for a region of the Portuguese coast

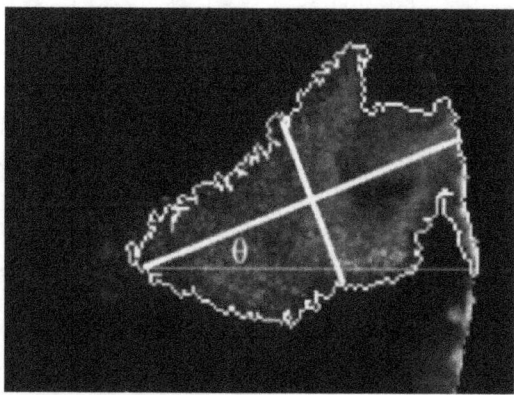

Figure 3. Illustration of the characterization of the river plume (2009-02-26) shape according to the adjusted ellipse (its major and minor axis) and the orientation of the river plume (adapted from [4])

Typically, outflow plumes are tracked in shelf water using density or salinity on account of the notably fresher composition of estuarine water. Unfortunately, there are no satellite-based remote-sensing platforms from which salinity can be directly measured. However, several studies have shown that these outflows carry large amounts of dissolved organic materials and suspended particles, which should allow plume events to be readily identified by remotely sensed optical images [77]. Kim et al. [78] related Chla concentration to salinity in the Changjiang plume area and presented the monthly summer plume area within a limited area during 1998–2007.

Hopkins et al. [79] used four satellite data products to examine the Sea Surface Temperature (SST), Sea Surface Salinity (SSS), Chla, and Mean Sea Level Anomaly (MSLA) fields in an area of the Angola Basin surrounding the Congo River mouth. Although it was not possible to extract a clear plume signature from the SST and MSLA alone, they provide useful supplementary understanding of the regions dynamics. Correlations between the SST, MSLA, Chla, and SSS help identify those areas persistently influenced by river input and those where variability is dominated by other processes.

4.2.3. Coral reef and Alga bloom

Coral reefs are one of the most biodiverse marine ecosystems on the planet. Worldwide, coral reef ecosystems are being increasingly threatened by sediment loads from river discharges, which in turn are influenced by changing rainfall patterns due to climate change and by growing human activity in their watersheds. Water turbidity and associated light attenuation are factors widely known to limit coral reef development. Coral reefs are generally limited to shallow and clear water with a mean water temperature of 18°C or higher, and are thus largely confined to the tropics [80]. Coral reef can be classified to show different forms of coral reef, dead coral, coral rubble, algal cover, sand, lagoons, different densities of seagrasses, etc. Several environmental variables have been shown to influence the biodiversity of a given habitat. Mapping such habitat variables could indicate the likely spatial distribution of biodiversity at a local scale and suggest priority areas for conservation, at least for the species for which habitat–biodiversity relationships have been identified. Remote-sensing technologies have been used to map coral reefs since the early days of Landsat program [81], and research into the use of remote-sensing technology continues with the advent of new sensors and data-processing methods [82]. The mapping of coral reefs and general bottom characteristics from satellites has become more accurate since high-resolution multispectral imagery became available [83, 84]. The development of hyperspectral instruments has also improved the degree to which accessory pigments can be used to separate detailed classes, and they have therefore enabled mapping of detailed classes while retaining satisfactory mapping accuracy [85, 86].

The work of Mumby et al. [85] reviews what can, might, and cannot be mapped using remote sensing, and not only covers aspects of reef structure and health but also discusses the diversity of physical environmental data such as temperature, winds, solar radiation, and water quality. Knudby et al. [87] reviewed coral reef biodiversity, the influence of habitat variables on its local spatial distribution, and the potential for remote sensing to produce maps of these habitat variables. Andréfouët et al. [88] present a review, where a new path is provided by following the diversity of units that have been mapped and characterized using high spatial resolution optical remote-sensing data for the main New Caledonian coral reef complexes and their individual reef-forming units. The combined examination of the different sources of data, and the exhaustive description of remotely sensed reef units, allows to a qualitative synoptic parallel to be drawn between the morphology of modern reefs and the contrasting patterns of reef growth, subsidence, and uplift rates occurring around New Caledonia. Hamel and Andréfouët [89] present a review about the use of very high resolution remote sensing for the management of coral reef fisheries. The rapid degradation of many reefs worldwide calls for more effective monitoring and predictions of the trajectories of coral reef habitats as they cross cycles of disturbance and recovery. Palandro et al. [90] used an 18-year (1984-2002) time series of Landsat 5/TM and 7/ETM+ images to assess changes in eight coral reef sites in the Florida Keys National Marine Sanctuary. A Mahalanobis distance classification was trained for four habitat classes. A detailed pixel-by-pixel examination of the spatial patterns across time suggests that the results range from ecologically plausible to unreliable due to spatial inconsistencies and/or improbable ecological successions.

Harmful Algal Blooms (HABs) phenomena are global and have been increasing in severity and extent, with many devastated implications. They cause eutrophic conditions, depleting oxygen levels needed for organic life, and limiting aquatic plant growth by reducing water transparency. HABs could be defined by an increase in the concentration of a phytoplankton species that has an adverse impact on the environment, with more serious implications when there is toxin production, but also with high biomass accumulation. HABs have been found to occur frequently in optically complex case-2 waters, such as in the Korean South Sea [91], East China Sea [92], Yellow Sea [91], Bohai Sea [93], Gulf of Mexico [94], among others. These blooms are dominated mostly by *Cochlodinium polykrikoides* (hereafter referred to as *C. polykrikoides*), *Alexandrium tamarense*, *Prorocentrum dentatum*, *Ceratium furca*, and *Karenia brevis*, causing massive mortalities of aquaculture fish and numerous ecological and health impacts since the last few decades. High concentrations of nutrients exported from agriculture or urban sprawl in coastal watersheds are also causing algal blooms in many estuaries and coastal waters [95]. Satellite detection and monitoring of HABs require methods/algorithms that have been developed mostly based on extensive in situ bio-optical observations from optically less complex oceanic waters and optical modeling of water properties. Remote-sensing bio-optical algorithms explore the optical properties (absorption, backscattering, and reflectance) of each water component (CDOM, TSM, and Chla) in order to establish equations that can indicate a relationship between the optical characteristics of each component and the total sensor signals. These relationships are generally obtained through empirical, semianalytical, or radiation transfer models, and require in situ data in order to validate the equations/models. However, this approach is only appropriate for case 1 waters. Several spectral band algorithms have been developed to overcome the limitation of the standard optical algorithms. One of the most common methods for identifying a HAB is to estimate the Chla concentration. More details about the use of remote-sensing techniques for detecting phytoplankton and mapping HABs could be found in Klemas [95]. A range of disciplines including biochemistry, physical oceanography, and geology can be brought together to improve the identification of HABs.

4.3. Wetland mapping and coastal hazards/vulnerability

The coastal zone represents a comparatively small but highly productive and extremely diverse system, with a variety of ecosystems. Remote sensing allows to quantitatively retrieving several parameters useful for produce multi-hazard and vulnerability maps [96], wetland mapping [97] and identify infestations of invasive plants [98]. Satellite remote sensors can map coastal ecosystems and their changes cost-effectively at appropriate scales and resolutions, minimizing the need for extensive field and ship measurements. Traditionally, the Landsat TM and ETM and SPOT data have been reliable data sources for wetlands mapping [99]. The current status of methodologies and the most innovative works will be described in the following. The final part of this section will also include a brief reference to beach monitoring/classification, due to its importance in coastal management.

Wetland health is strongly impacted by runoff from land and its use within the same watershed. To study the impact of land runoff on estuarine and wetland ecosystems, a combination of models is frequently used, including watershed models, hydrodynamic models, and water-quality models [100]. The availability of high spatial and spectral resolution satellite data has significantly improved the capacity for mapping salt marshes and other coastal ecosystems [101]. Major plant species within a complex, heterogeneous tidal marsh have been classified using multitemporal, high-resolution images. Hyperspectral data have also been used for mapping coastal wetlands. The advantages and problems associated with hyperspectral mapping have been clearly demonstrated by Hirano et al. [102]. A number of techniques have been developed for mapping wetlands and even identifying wetland types and plant species [99, 103, 104]. To identify long-term trends and short-term variations, such as the impact of rising sea levels and hurricanes on wetlands, one needs to analyze time-series of remotely sensed imagery [105, 106]. Submerged aquatic vegetation is an important part of wetland and coastal ecosystems, playing a major role in the ecological functions of these habitats. Alga bloom and coral reefs have been discussed in section 3.3. The classification of Land Use and Land Cover (LULC) in delta regions was also the subject of several works. For instance, Fan et al. [107] investigated LULC in the Pear River Delta (China) using Landsat TM and ETM+ images and employed ML classifier. El-Kawy et al. [108] applied a supervised classification (ML) to four Landsat images (TM and ETM+) collected between 1984 and 2009 that provided recent and historical LULC conditions for the western Nile delta. The LULC mapping accuracy of 96% indicates that the integration of visual interpretation with the supervised classification of remote-sensing imagery is an effective method for the identification of changes in LULC. More recently, Tran et al. [109] presented a study where the main objective was to assess the spatiotemporal dynamics of LULC changes in the lower Mekong Delta (Vietnam) over the last 40 years. LULC change dynamics are derived from Landsat and SPOT satellite imagery between 1973 and 2011.

Vulnerability can be defined as the degree to which a person, community, or a system is likely to experience harm due to exposure to an external stress. Vulnerability also encompasses the idea of response and coping, since it is determined by the potential of a community to react and withstand a disaster [110]. A Multi-Hazard Vulnerability Map (MHVM) incorporates vulnerability in understanding the risk due to a hazard. Mahendra et al. [96] present a study that aims developing a methodology for assessing the multi-hazard vulnerability and gather quantitative estimate on the spatial extent of the inundation caused by composite hazards in Tamil Nadu state in the Bay of Bengal (India). The parameters used in this study were: shoreline change rate, sea level change rate, historical storm surges, and the high-resolution topography. Data from Landsat MSS, TM, and ETM and QuickBird were used to extract some parameters and, afterward, generate the hazard and risk maps. Risk maps and evacuation routes are generated by imbibing land use, transport, and structural information. Scientific study of the natural hazards and coastal processes of the Indian coast has assumed greater significance after the December 2004 tsunami because the country learned lessons on the impact of natural hazards in terms of high damage potential for life, property, and the environment. Several works were published related to this topic. Römer et al. [111] presents a case study that focuses on a local assessment of tsunami hazard and vulnerability, including the socioeconomic and ecological components. High-resolution optical data (IKONOS-2) were employed to create basic geo-data including LULC, to provide input data for the hazard and vulnerability assessment. Results show that the main potential of applying remote-sensing techniques and data derives from a synergistic combination with other types of data. Kumar et al. [27] develop a coastal vulnerability index for the maritime state of Orissa (India), using eight relative risk variables. Ortho-rectified Landsat MSS and TM images covering the Orissa coastline (India) for the years 1970, 1980, and 2000 were used to digitize the shoreline. Indian Remote Sensing Satellite (IRS) P6 Linear Imaging Self-scanning Sensor-IV (LISS-IV) was used to extract the coastal geomorphology. Zones of vulnerability to coastal natural hazards of different magnitude are identified.

Beach morphological classification was mainly based on in situ data (wave, tidal, and sediment parameters). However, parameters such as those are usually unavailable for several coastal areas. Optical remote sensing is a very powerful tool for beach monitoring/classification, since it allows identification and classification of beach morphologies. Teodoro et al. [112] applied a pixel-based (supervised or unsupervised) and region-based (object-oriented classification) classification to high-resolution data (aerial photographs and IKONOS-2 image) in order to identify, measure, and classify beach features/patterns and further classify the beach extension considered (Northwest coast of Portugal). Thereafter, in order to implement an automatic beach patterns extraction methodology, Teodoro et al. [113] present a new approach based on Principal Components Analysis and Histogram segmentation (PCAH) aiming to identify and analyze morphological features and hydrodynamic patterns, also applied to aerial photographs and IKONOS-2 image. More recently, Teodoro [114] applied data-mining techniques, particularly ANN and Decision Trees (DT), to the same image in order to identify and classify beach features and their geographic patterns. Teodoro [114] concludes that the use of ANNs and DTs for beach classification from optical remotely sensed data resulted in an increased classification accuracy when compared with traditional classification methods, as shown in Fig. 4. The results of this work should be used as an input in beach classification models, in sediment budget estimation, and also in the identification/characterization of rip currents and bars (location, spacing, persistence, size, and strength).

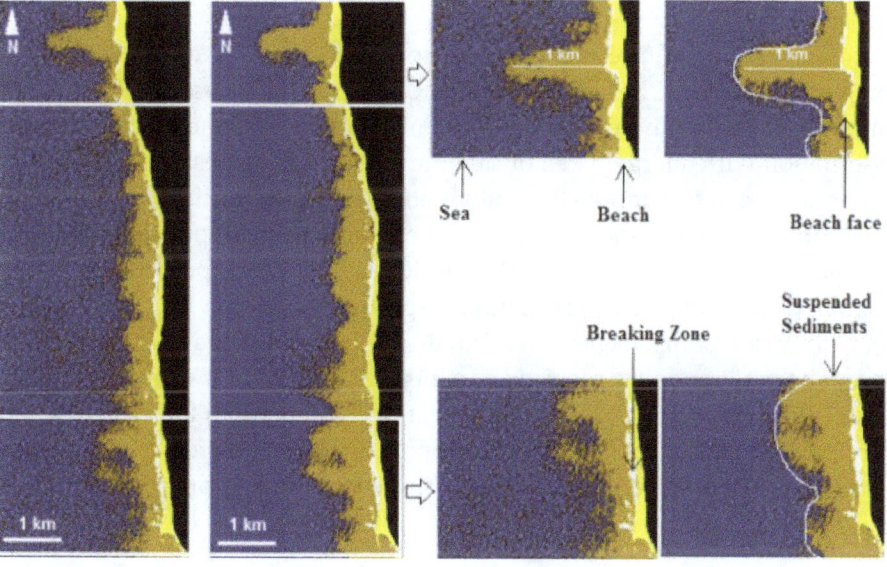

Figure 4. Beach patterns/forms identification and two zoomed areas obtained through (a) DT with pruning and (b) ANN (adapted from Teodoro [114])

Traditionally, the Landsat TM and ETM and SPOT satellite have been reliable data sources for wetlands mapping. However, in recent years the use of high spatial resolution data and hyperspectral data has become quite popular. In the vulnerability and hazards studies, different optical satellite data were commonly used to extract some parameters (e.g., shoreline change rate, land use) essential to hazard and risk maps generation. The use of high spatial resolution data is crucial. All these approaches are the key for a correct and efficient management of coastal environments.

4.4. Bathymetry

Bathymetric information is of crucial importance in coastal areas, such as in estuarine areas, which often exhibit a high population density, and vulnerable natural ecosystems. Optical remote sensing offers a cost-effective alternative to echo sounding and bathymetric LiDAR techniques for deriving bathymetric estimates in shallow coastal and inland waters [115- 117]. Images from optical remote sensors possess attractive properties for bathymetric mapping, including synoptic coverage of water surface areas, wide availability for most geographical regions, and relatively low cost [118]. The availability of optical high-resolution satellites, such as IKONOS, QuickBird, and WorldView, has renewed interest in applying optical remote-sensing techniques to the retrieval of bathymetric information for shallow coastal and inland waters, due to their high spatial resolution and enhanced water penetration capability. In this context, several inversion algorithms and models have been proposed in the literature for retrieving bottom depth estimates from multispectral remote-sensing imagery [115, 116, 118-120]. The simplest method of retrieving water depth from single-band remote-sensing imagery was first proposed by Lyzenga [121]. Later, Lyzenga [122, 123] derived a log-linear inversion model for inverting multispectral imagery to water depth. This inversion model uses the linear logarithmic-transformed multispectral remote-sensing data as the predictors to estimate water depth.

Minghelli-Roman et al. [124] present a comparison of bathymetric estimation using different satellite images (Quickbird, ETM, Hyperion, MERIS) in coastal seawaters. The aim of this study was to compare, for one bathymetric estimation method and one mesotrophic site, the results of depth estimation with a large panel of satellite and aerial images. For each image, the pair of spectral bands chosen to compute the bathymetry has been optimized. This comparison was discussed, in order to identify the influence of image parameters (spectral bands, SNR, spatial resolution, and quantization) on the bathymetric results and to propose the most adapted image parameters for bathymetric estimation. Regarding the depth RMSE errors obtained, no sensor seems to be the perfect sensor to estimate bathymetry. Regarding the spectral config-uration, three spectral bands are required to generate the mask on water: the first in the blue-green domain; the second in the green domain; and a final band in the near-infrared domain. The atmospheric correction has to be efficient because a strong diffusion operates in the blue domain. A very high resolution such as Quickbird's is not necessary, but a lower resolution than 30 m induces mixed pixels on the shore and then degrades the estimation in shallow waters.

Teodoro et al. [125] propose a model for the estimation of depth based on Principal Component Analysis (PCA) of an IKONOS-2 image, for the Douro River estuary (Porto, Portugal). Subsequently, Teodoro et al. [117] proposed alternative univariate and bivariate models for the same IKONOS-2 image based on PCA and Independent Component Analysis (ICA). The PCA is the standard method for separating mixed signals. Such analysis provides signals that are linearly uncorrelated. Although the separated signals are uncorrelated they could still be depended, i.e., nonlinear correlation remains. The ICA was developed to investigate such data. The results obtained were compared with the bathymetric estimation through PCA. Best univariate ICA-based model allowed to estimate depth with a mean error that outperforming the best PCA based univariate model results, even with the first PCA component explains 80% of data variance. With bivariate models the results improved.

Kanno et al. [116] proposed a method that combines a spatial interpolation method based on nonparametric regression and Lyzenga et al. [115] method on a statistical basis. A multispectral image of QuickBird of a coral reef site along Ishigaki City (Japan) was used in this approach. This method is based on a semiparametric regression model that consists of a parametric imagery-based term and a nonparametric spatial interpolation term that complement one another. An accuracy comparison in a test site showed that this new method is more accurate than either of the existing methods when sufficient training data are available and far more accurate than the spatial interpolation method when the training data are scarce.

Su et al. [118] propose a geographically adaptive inversion model for improving bathymetric retrieval in complex and heterogeneous marine environments for Hawaiian Islands. By using IKONOS-2 and Landsat ETM+ images, they demonstrated that regionally and locally calibrated inversion models can effectively address the problems introduced by spatial heterogeneity in water quality and bottom type, and provide significantly improved bathymetric estimates for more complex coastal waters.

More recently, Eugenio et al. [126] presented an optimal atmospheric correction model, as well as an improved algorithm for sunglint removal based on combined physical and image-processing techniques. The spectral capabilities of World View-2 multispectral imagery (for Granadilla in Tenerife Island and Corralejo in Fuerteventura Island) was exploited for bathymetry retrieval. Using the radiative model to compute bathymetry has yielded good results and allowed to improve the outcome of the ratio algorithm as it considers the physical phenomena of water absorption and backscattering and the relationship between the seafloor albedo, its depth, and the water IOPs. The accuracy of the proposed bathymetry retrieval algorithm output for each coastal area image was assessed with a scatter plot of the algorithm output versus acoustic field data.

In the recent years, several methods based in inversion algorithms and radiative models have been proposed in the literature for retrieving bottom depth from optical remote-sensing imagery. Other approaches have also been tested mainly based in statistical methods. The use of high-resolution optical images seems to improve the accuracy of depth estimation. However, several problems related to atmospheric conditions, SNR, and seafloor contributions are yet to be resolved. There is still a long way to go in using this type of data to estimate the depth for coastal environments through optical remote-sensing data.

5. Conclusions

Different optical satellite data and different methodologies could be used to monitor the coastal environment. There is not an ideal sensor, or an effective technique/algorithm that can be applied to all the coastal environments components/parameters. Depending on what parameter or element is being studied, the selected sensor should have the best characteristics (spatial, spectral, radiometric, and temporal resolution) for the objective proposed. The optimal spatial resolution for the assessment of coastal ecosystems is not consensual. Despite the high spatial resolution images that provide more detail, for several studies low or moderate spatial resolution is enough. Moreover, the low spatial coverage of the high spatial resolution images could be a limiting factor for regional or global studies. The recent developments of hyper-spectral sensors that provide very high spectral resolutions introduce a new scenario in this field, allowing, for instance, the development of bio-optical algorithms, more adequate for coastal zones environments. The temporal resolution also depends on the objectives of the research. Various image-processing techniques have been applied to the satellite images in order to study the coastal environment. These techniques differ depending on the subject of study. In the shoreline change detection, beyond visual interpretation, several image segmen-tation and image classification algorithms are used to identify and detect the evolution of the coastline. Also, several types of algorithms are employed in the quantification of water constituents. A variety of parameters, including Chla, TSM, turbidity, and aCDOM, can be estimated. For instance, in the estimation of Chla, the 700/670 nm ratio reflectance (for high-biomass waters) has been widely used. Alternatively, more complex algorithms, such as ANN, can be employed. Many TSM models are based on empirical methods. However, other algorithms, such as ANN, can also be applied to retrieve the TSM concentration. The identi-fication and monitoring of river plumes can be done considering the water constituents (TSM, salinity, Chla) or applying segmentation and classification algorithms that allow identification of the plume boundaries. The detection and monitoring of HABs require algorithms that have been developed mostly based on extensive in situ bio-optical observations from optically less complex oceanic waters and optical modeling of water properties. Remote-sensing bio-optical algorithms explore the optical properties of each water component. A number of techniques have also been developed for mapping wetlands hazards/vulnerability. When LULC is required, different image classification algorithms can be used. Other algorithms, such PCA, DT, and ANN, can also be used, for instance, in the identification of beach patterns. In the bathymetric estimation, beyond the inversion algorithms and radiative models widely applied, statistical algorithms, such as PCA and ICA can also be used to estimate the depth for estuarine areas. Several advances were discussed related to the recent availability of data from new sensors and hyperspectral data. In short, the assessment of the Sentinel-2 data will improve coastal environment monitoring programs. The elimination of the degree of uncer-tainty in some procedures should be a priority. There are available at present, a lot of robust, well-tested algorithms that allow quantification and accurate estimation of several parameters. The major challenge still is to have remote-sensing techniques adopted as a routine tool in assessment of change in the coastal zone. Continuing research is required into the techniques employed for assessing change in the coastal environment.

Author details

Ana C. Teodoro[*]

Address all correspondence to: amteodor@fc.up.pt

Earth Sciences Institute (ICT) and Department of Geosciences, Environment and Land Planning, Faculty of Sciences, University of Porto, Porto, Portugal

References

[1] Cracknell AP: Remote sensing techniques in estuaries and coastal zones – an update. *Int J Remote Sens*. 1999;20(3):485-496.

[2] Malthus TJ, Mumby PJ: Remote sensing of the coastal zone: an overview and priorities for future research. *Int J Remote Sens*. 2003;24(13):2805-2815.

[3] Klemas V: Remote sensing of emergent and submerged wetlands: an overview. *Int J Remote Sens*. 2013;34(18):6286-6320.

[4] Kerekes JP. Optical sensor technology. In: *The SAGE Handbook of Remote Sensing*. Warner TA, Nelis MD, Foody GM, editors. SAGE Publications; 2009. pp. 95-107.

[5] Mouw CB, Greb S, Aurin D, DiGiacomo P, Lee Z, Twardowski M, Binding C, Hu C, Ma R, Moore T, Moses W, Craig S: Aquatic color radiometry remote sensing of coastal and inland waters: challenges and recommendations for future satellite missions. *Remote Sens Environ*. 2015;160:15-30.

[6] Gonçalves H, Teodoro AC, Almeida H: Identification, characterization and analysis of the Douro River Plume from MERIS data. *IEEE J Select Topics Appl Earth Observ Remote Sens*. 2012;5(5):1553-1563.

[7] Feng L, Hu C, Chen X, Tian L, Chen L: Human induced turbidity changes in Poyang Lake between 2000 and 2010: Observations from MODIS. *J Geophys Res*. 2012;117:C07006.

[8] Palmer SCJ, Kutser T, Hunter PD: Remote sensing of inland waters: Challenges, progress and future directions. *Remote Sens Environ*. 2015;157:1-8.

[9] Davranche A, Lefebvre G, Poulin B: Wetland monitoring using classification trees and SPOT-5 seasonal time series. *Remote Sens Environ*. 2010;114(3):552-562.

[10] Santillan J, Makinano M, Paringit E: Integrated Landsat image analysis and hydrologic modeling to detect impacts of 25-Year land-cover change on surface runoff in a Philippine watershed. *Remote Sens*. 2011;3(6):1067-1087.

[11] Hellweger FL, Schlosser P, Lall U, Weissel JK: Use of satellite imagery for water quality studies in New York Harbor. *Estuar Coast Shelf Sci*. 2004;61(3):437-448.

[12] Teodoro AC, Marçal ARS, Veloso-Gomes F: Correlation analysis of water wave reflectance and local TSM concentrations in the breaking zone, using remote sensing techniques. *J Coast Res*. 2007;23(6):1491-1497.

[13] Teodoro AC, Veloso-Gomes F, Goncalves H: Retrieving TSM concentration from multispectral satellite data by multiple regression and artificial neural networks. *IEEE Trans Geosci Remote Sens*. 2007;45(5):1342-1350.

[14] Teodoro AC, Veloso-Gomes F: Quantification of the Total Suspended Matter concentration around the sea breaking zone from in situ measurements and TERRA/ASTER data. *Mar Geores Geotechnol*. 2007;25(2):67-80.

[15] Ouillon S, Douillet P, Petrenko A: Optical algorithms at satellite wavelengths for total suspended matter in tropical coastal waters. *Sensors*. 2008;8:4165-4185.

[16] Ruelland D, Dezetter A, Puech C, Ardoin-Bardin S: Long-term monitoring of land cover changes based on Landsat imagery to improve hydrological modelling in West Africa. *Int J Remote Sens*. 2008;29(12):3533-3551.

[17] Drusch M, Del Bello U, Carlier S, Colin O, Fernandez V, Gascon F, Hoersch B, Isola C, Laberinti P, Martimort P, Meygret A, Spoto F, Sy O, Marchese F, Bargellini P: Sentinel-2: ESA's optical high-resolution mission for GMES operational services. *Remote Sens Environ*. 2012;120:25-36.

[18] Vitousek PM, Mooney HA, Lubchenco J, Melillo JM: Human domination of earth's ecosystems. Science. 1997;277(5325):494-499.

[19] IPCC, 2007: Climate Change 2007: Impacts, Adaptation and Vulnerability. Contribution of Working Group II to the Fourth Assessment Report of the Intergovernmental Panel on Climate Change, M.L. Parry, O.F. Canziani, J.P. Palutikof, P.J. van der Linden and C.E.

[20] Hanson, Eds., Cambridge University Press, Cambridge, UK, 976pp.

[21] Dolan R, Fenster MS, Holme SJ: Temporal analysis of shoreline recession and accretion. *J Coastal Res*. 1991; 7:723-744.

[22] Boak EH, Turner IL: Shoreline definition and detection: a review. *J Coastal Res*. 2005; 21:688-703.

[23] Gens R: Remote sensing of coastlines: detection, extraction and monitoring. *Int J Remote Sens*. 2010;31(7): 1819-1836.

[24] Guariglia A, Buonamassa A, Losurdo A, Saladino R, Trivigno ML, Zaccagnino A, Colangelo A: A multisource approach for coastline mapping and identification of shoreline changes. *Ann Geophys*. 2006;49:295-304.

[25] Ekercin S: Coastline change assessment at the Aegean Sea Coasts in Turkey using multitemporal Landsat imagery. *J Coastal Res*. 2007;23(3):691-698.

[26] Maiti S, Bhattacharya AK: Shoreline change analysis and its application to prediction: A remote sensing and statistics based approach. *Mar Geol*. 2009; 257:11-23.

[27] Kuleli T, Guneroglu A, Karsli F, Dihkan M: Automatic detection of shoreline change on coastal Ramsar wetlands of Turkey. *Ocean Engin*. 2011;38:1141-1149.

[28] Kumar TS, Mahendra RS, Nayak S, Radhakrishnan K, Sahu KC: Coastal vulnerability assessment for Orissa State, East Coast of India. *J Coastal Res*. 2010;26(3):523-534.

[29] Wang CY, Zhang J, Ma Y: Coastline interpretation from multispectral remote sensing images using an association rule algorithm. *Int J Remote Sens*. 2010;31(24):6409-6423.

[30] Teodoro AC, Gonçalves H: A semi-automatic approach for the extraction of sandy bodies (sand spits) from IKONOS-2 data. *IEEE J Select Topics Appl Earth Observ Remote Sens*. 2012;5(2):634-642.

[31] Pardo-Pascual JE, Almonacid-Caballer J, Ruiz LA, Palomar-Vázquez J: Automatic extraction of shorelines from Landsat TM and ETM+ multi-temporal images with sub-pixel precision. *Remote Sens Environ*. 2012;123:1-11.

[32] García-Rubio G., Huntley D, Russell P: Evaluating shoreline identification using optical satellite images. *Mar Geol*. 2015;359:96-105.

[33] Myint SW, Walker ND: Quantification of surface suspended sediments along a river dominated coast with NOAA AVHRR and SeaWiFS measurements: Louisiana, USA. *Int J Remote Sens*. 2002;23(16):3229-3249.

[34] Matthews MW: A current review of empirical procedures of remote sensing in inland and near-coastal transitional waters. *Int J Remote Sens*. 2011;32(21):6855-6899.

[35] Moses WJ, Gitelson AA, Berdnikov S, Povazhnyy V: Estimation of chlorophyll-a concentration in case II waters using MODIS and MERIS data – successes and challenges. *Environ Res Lett*. 2009;4:045005.

[36] Kiefer I, Odermatt D, Anneville O, Wuest A, Bouffard D: Application of remote sensing for the optimization of in-situ sampling for monitoring of phytoplankton abundance in a large lake. *Sci Total Environ*. 2015;527:493-506 2015.

[37] Gitelson AA, Gurli D, Moses WJ, Barrow, T: A bio-optical algorithm for the remote estimation of the chlorophyll-a concentration in case 2 waters. *Environ Res Lett*. 2009;4:045003.

[38] Le C, Li Y, Zha Y, Sun D, Huang C, Lu H: A four-band semi-analytical model for estimating chlorophyll a in highly turbid lakes: the case of Taihu Lake, China. *Remote Sens Environ*. 2009;113:1175-1182.

[39] Gitelson A, Mayo M, Yacobi YZ, Parparov A, Berman T: The use of highspectral- resolution radiometer data for detection of low chlorophyll concentrations in Lake Kinneret. *J Plankton Res*. 1994;16:993-1002.

[40] Ye HB, Chen CQ, Sun ZH, Tang SL, Song XY, Yang CY, Tian LQ, Liu, FF: Estimation of the primary productivity in Pearl River Estuary using MODIS data. *Estuaries Coasts*. 2015;38(2):506-518.

[41] Giardino C, Candiani G, Zilioli E: Detecting chlorophyll-a in Lake Garda using TOA MERIS radiances. *Photogram Engin Remote Sens*. 2005;71:1045-1051.

[42] Duan HT, Zhang YZ, Zhan B, Song KS, Wang ZM: Assessment of chlorophyll-a concentration and trophic state for Lake Chagan using Landsat TM and field spectral data. *Environ Monitor Assess*. 2007;129(1-3):295-308.

[43] Su YF, et al: A multivariate model for coastal water quality mapping using satellite remote sensing images. *Sensors*. 2008;8(10):6321-6339.

[44] Ormeci C, Sertel E, Sarikaya O: Determination of chlorophyll-a amount in Golden Horn, Istanbul, Turkey using IKONOS and in situ data. *Environ Monitor Assess*. 2009;155(1-4):83-90.

[45] Wu M, Zhang W, Wang X, Luo D: Application of MODIS satellite data in monitoring water quality parameters of Chaohu Lake in China. *Environ Monitor Assess*. 2009;148:255-264.

[46] Tyler AN, Svab E, Preston T, Presing M, Kovacs WA: Remote sensing of the water quality of shallow lakes: A mixture modelling approach to quantifying phytoplankton in water characterized by high-suspended sediment. *Int J Remote Sens*. 2006;27(8): 1521-1537.

[47] Oyama Y, Matsushita B, Fukushima T, Matsushige K, Imai A: Application of spectral decomposition algorithm for mapping water quality in a turbid lake (Lake Kasumigaura, Japan) from Landsat TM data. *ISPRS J Photogram Remote Sens*. 2009;64:73-85.

[48] Doeffer R, Fischer J, Stössel M, Brockman C: Analysis of Thematic Mapper data for studying the suspended matter distribution in the coastal area of the German bight (North Sea). *Remote Sens Environ*. 1989;28:61-73.

[49] Nechad B, Ruddick KG, Park Y: Calibration and validation of a generic multisensor algorithm for mapping of total suspended matter in turbid waters. *Remote Sens Environ*. 2010;114:854-866.

[50] Kong, JL et al.: A semi-analytical model for remote sensing retrieval of suspended sediment concentration in the Gulf of Bohai, China. *Remote Sens*. 2015;7(5):5373-5397.

[51] Kutser T, Arst H, Mäekivi S, Kallaste, K: Estimation of the water quality of the Baltic Sea and lakes in Estonia and Finland by passive optical remote sensing measurements on board vessel. *Lakes Reserv: Res Manag*. 1998;3:53-66.

[52] Bowers D, Harker G, Smith P, Tett, P: Optical properties of a region of freshwater influence (the Clyde Sea). *Estuarine Coastal Shelf Sci*. 2000;50:717-726.

[53] Doxaran D, Cherukuru R, Lavender S: Use of reflectance band ratios to estimate suspended and dissolved matter concentrations in estuarine waters. *Int JRemote Sens*. 2005;26:1763-1770.

[54] D'sa EJ, Miller RL: Bio-optical properties in waters influenced by the Mississippi River during low flow conditions. *Remote Sens Environ*. 2003;84:538-549.

[55] Koponen S, Attila J, Pulliainen J, Kallio K, Pyhälahti T, Lindfors A, Rasmus K, Hallikainen M: A case study of airborne and satellite remote sensing of a spring bloom event in the Gulf of Finland. *Continental Shelf Res.* 2007;27:228-244.

[56] Kutser T, Paavel B, Metsamaa L: Mapping coloured dissolved organic matter concentration in coastal waters. *Int J Remote Sens.* 2009;30:5843-5849.

[57] Loisel H, Vantrepotte V, Dessailly D, Meriaux X: Assessment of the colored dissolved organic matter in coastal waters from ocean color remote sensing. *Optics Express.* 2014;22(11):13109-13124.

[58] Kirk JTO: Optical water quality: What does it mean and how should we measure it?. *J Water Poll Contr Fed.* 1988;60:194-197.

[59] Sun D, Li Y, Wang Q, Gao J, Lv H, Le C, Huang C: Light scattering properties and their relation to the biogeochemical composition of turbid productive waters: A case study of Lake Taihu. *Appl Optics.* 2009;48:1979-1989.

[60] Gitelson AA, Dall'Olmo G, Moses W, Rundquist DC, Barrow T, Fisher TR, Gurlin D, Holz J: A simple semi-analytical model for remote estimation of chlorophyll-a in turbid waters: Validation. *Remote Sens Environ.* 2008;112(9):3582-3593.

[61] Barnes BB, Hu C, Holekamp KL, Blonski S, Spiering BA, Palandro D, et al. Use of Landsat data to track historical water quality changes in Florida Keys marine environments. *Remote Sens Environ.* 2014;140: 485-496.

[62] Hommersom A, Wernand MR, Peters S, de Boer J: A review on substances and processes relevant for optical remote sensing of extremely turbid marine areas, with a focus on the Wadden Sea. *Helgol Mar Res.* 2010;64:75-92.

[63] Dekker AG, Vos RJ, Peters SWM: Comparison of remote sensing data, model results and in situ data for total suspended matter (TSM) in the southern Frisian lakes. *Sci Total Environ.* 2001;268:197-214.

[64] Kaufman, YJ: Solution of the equation of radiative-transfer for remote-sensing over nonuniform surface reflectivity. *J Geophys Res – Oceans Atmos.* 1982;87(NC6): 4137-4147.

[65] Vermote EF, et al.: Second simulation of the satellite signal in the solar spectrum, 6S: An overview. *IEEE Trans Geosci Remote Sens.* 1997;35(3):675-686.

[66] Defoin-Platel M, Chami M: How ambiguous is the inverse problem of ocean color in coastal waters? *J Geophys Res.* 2007;112:C03004.1-C03004.16.

[67] Mélin F, Vantrepotte V: How optically diverse is the coastal ocean? *Remote Sens Environ.* 2015;160:235-251.

[68] Mertes LAK, Warrick JA: Measuring flood output from 110 coastal watersheds in California with field measurements and SeaWiFS. *Geology.* 2001;29:659-662.

[69] Otero P, Ruiz-Villarreal M, Peliz A: River plume fronts off NW Iberia from satellite observations and model data. *ICES J Mar Sci: J du Conseil*. 2009;66(9):1853-1864.

[70] Son YB, Gardner WD, Richardson MJ, Ishizaka J, Ryu JH, Kim SH, Lee SH:Tracing offshore low-salinity plumes in the Northeastern Gulf of Mexico during the summer season by use of multispectral remote-sensing. *J Oceanography*. 2012;68(5):743-760.

[71] Fernández-Nóvoa D, Mendes R, deCastro M, Dias JM, Sánchez-Arcilla A, Gómez-Gesteira M: Analysis of the influence of river discharge and wind on the Ebro turbid plume using MODIS-Aqua and MODIS-Terra data. *J Mar Sys*. 2015;142:40–46.

[72] Teodoro AC, Gonçalves H, Veloso-Gomes F, Gonçalves JA: Modelling of the Douro river plume size, obtained through image segmentation of MERIS data. *IEEE Geosci Remote Sens Lett*. 2009;6(1):87-91.

[73] Jiang L, Yan XH, Klemas V: Remote sensing for the identification of coastal plumes: case studies of Delaware Bay. *Int J Remote Sens*. 2009;30(8):2033-2048.

[74] Guneroglu A, Karsli F, Dihkan M: Automatic detection of coastal plumes using Landsat TM/ETM plus images. *Int J Remote Sens*. 2013;34(13):4702-4714.

[75] Zhu WN, Yu Q: Inversion of chromophoric dissolved organic matter (CDOM) from EO-1 Hyperion imagery for turbid estuarine and coastal waters. *IEEE Trans Geosci Remote Sens*. 2013;51(6): 3286-3298.

[76] Lihan T, Saitoh S, Lida T, Hirawake T, Lida K: Satellite-measured temporal and spatial variability of the Tokachi River plume. *Estuarine Coastal Shelf Sci*. 2008;78:237-249.

[77] Dzwonkowski B, Yan X: Tracking of a Chesapeake Bay estuarine outflow plume with satellite-based ocean color data. *Continental Shelf Res*. 2005;25:1942-1958.

[78] Kim HC, et al.: Distribution of Changjiang diluted water detected by satellite chlorophyll-a and its inter-annual variation during 1998–2007. *J Oceanography*. 2009;65:129-135.

[79] Hopkins J, Lucas M, Dufau C, Sutton M, Stum J, Lauret O, Channelliere C: Detection and variability of the Congo River plume from satellite derived sea surface temperature, salinity, ocean colour and sea level. *Remote Sens Environ*. 2015;139:365-385.

[80] Yonge CM. *The Biology of Reef Building Corals*. London: British Museum, 1940. 353 p.

[81] Smith VE, Rogers RH, Reed LE: Automated mapping and inventory of Great Barrier Reef zonation with Landsat. *Oceans*. 1975;7:775-780.

[82] Kutser T, Miller I, Jupp D: Mapping coral reef benthic substrates using hyperspectral space-borne images and spectral libraries. *Estuarine Coastal Shelf Sci*. 2006;70:449-460.

[83] Purkis SJ: A "Reef-Up" approach to classifying coral habitats from IKONOS imagery. *IEEE Trans Geosci Remote Sens*. 2005;43:1375-1390.

[84] Mishra D, Narumalani S, Rundquist D, Lawson M: Benthic habitat mapping in tropical marine environments using QuickBird multispectral data. *Photogram Engin Remote Sens*. 2006;72:1037-1048.

[85] Mumby PJ, Skirving W, Strong AE, Hardy JT, LeDrew EF, Hochberg EJ, Stumpf RP, David LT: Remote sensing of coral reefs and their physical environment. *Mar Poll Bull*. 2004;48(3-4):219-228.

[86] Kutser T, Dekker AG, Skirving W: Modeling spectral discrimination of Great Barrier Reef benthic communities by remote sensing instruments. *Limnol Oceanography*. 2003;48:497-510.

[87] Knudby A, LeDrew E, Newman C: Progress in the use of remote sensing for coral reef biodiversity studies. *Progr Phys Geography*. 2007;31(4):421-434.

[88] Andréfouët S, Cabioch G, Flamand B, Pelletier B: A reappraisal of the diversity of geomorphological and genetic processes of New Caledonian coral reefs: a synthesis from optical remote sensing, coring and acoustic multibeam observations. *Coral Reefs*. 2009;28(3):691-707.

[89] Hamel MA, Andréfouët S: Using very high resolution remote sensing for the management of coral reef fisheries: Review and perspectives. *Mar Poll Bull*. 2010;60(9): 1397-1405.

[90] Palandro DA, Andrefouet S, Hu C, Hallock P, Muller-Karger FE, Dustan P, Callahan MK, Kranenburg C, Beaver CR: Quantification of two decades of shallow-water coral reef habitat decline in the Florida Keys National Marine Sanctuary using Landsat data (1984-2002). *Remote Sens Environ*. 2008;112(8):3388-3399.

[91] Ahn YH, Shanmugam P, Ryu JH, Jeong JC: Satellite detection of harmful algal bloom occurrences in Korean waters. *Harmful Algae News*. 2006; 213-131.

[92] Gao X, Song J: Phytoplankton distributions and their relationship with the environment in the Changjiang Estuary, China. *Mar Poll Bull*. 2005;50:327-335.

[93] Tang DL, Kawamura H, Oh IS, Baker J: Satellite evidence of harmful algal blooms and related oceanographic features in the Bohai Sea during autumn 1998. *Adv Space Res*. 2006;37:681-689.

[94] Tomlinson M., Wynne TT, Stumpf RP: An evaluation of remote sensing techniques for enhanced detection of the toxic dinoflagellate, Karenia brevis. *Remote Sens Environ*. 2009;113:598-609.

[95] Klemas V: Remote sensing of algal blooms: An overview with case studies. *J Coastal Res*. 2012;28(1A):34-43.

[96] Mahendra RS, Mohanty PC, Bisoyi H, Srinivas T, Kumar, Nayak, S: Assessment and management of coastal multi-hazard vulnerability along the Cuddalore Villupuram, east coast of India using geospatial techniques. *Ocean Coastal Manag*. 2011;54: 302-311.

[97]   Gilmore MS, Wilson EH, Barrett N, Civco DL, Prisloe S, Hurd JD, Chadwick C: Inte-
       grating multitemporal spectral and structural information to map wetland vegetation
       in a Lower Connecticut River Tidal Marsh. *Remote Sens Environ.* 2008;112:4048-4060.

[98]   Cavalli RM, Laneve L, Fusilli S, Pignatti F, Santini: Remote sensing water observation
       for supporting Lake Victoria weed management. *J Environ Manag.* 2009;90(7):
       2199-2211.

[99]   Klemas V: Remote sensing of wetlands: Case studies comparing practical techniques.
       J Coastal Res. 2011;27: 418-427.

[100] Li M, L Zhong WC, Boicourt S, Zhang, Zhang D: Hurricane-induced stratification
       and destratification in a partially-mixed estuary. *J Mar Res.* 2007;65:169-192.

[101] Laba M, Downs R, Smith S, Welsh S, Neider C, White S, Richmond M, Philpot W,
       Baveye P: Mapping invasive wetland plants in the Hudson River National Estuarine
       Research Reserve using Quickbird satellite imagery. *Remote Sens Environ.* 2008;112:
       286-300.

[102] Hirano A, Madden M, Welch R: Hyperspectral image data for mapping wetland veg-
       etation. *Wetlands.* 2003;23:436-448.

[103] Yang X. *Remote Sensing and Geospatial Technologies for Coastal Ecosystem Assessment and
       Management.* Berlin: Springer-Verlag; 2009.

[104] Heumann BW: Satellite remote sensing of mangrove forests: Recent advances and fu-
       ture opportunities. *Progr Phys Geography.* 2011;35:87-108.

[105] Baker C, Lawrence RL, Montagne C, Patten D: Change detection of wetland ecosys-
       tems using Landsat imagery and change vector analysis. *Wetlands.* 2007;27:610-619.

[106] Ramsey E, Rangoonwala A. Mapping the onset and progression of marsh dieback.
       In: *Remote Sensing of Coastal Environment*, J. Wang, editor. Boca Raton, FL: CRC Press;
       2010. pp. 123-150.

[107] Fan F, Wang Y, Wang Z: Temporal and spatial change detecting (1998–2003) and pre-
       dicting of land use and land cover in Core corridor of Pearl River Delta (China) by
       using TM and ETM+ images. *Environ Monitoring Assess.* 2008;137:127-147.

[108] Abd El-Kawy OR, Rød JK, Ismail HA, Suliman AS: Land use and land cover change
       detection in the western Nile delta of Egypt using remote sensing data. *Appl Geogra-
       phy.* 2011;31:483-494.

[109] Tran H, Tran T, Kervyn M: Dynamics of land cover/land use changes in the Mekong
       Delta, 1973-2011: A remote sensing analysis of the Tran Van Thoi District, Ca Mau
       Province, Vietnam. *Remote Sens.* 2015;7:2899-2925.

[110] Kumpulainen S. Vulnerability concepts in hazard and risk assessment. Natural and
       technological hazards and risks affecting the spatial development of European re-

gions. In: Schmidt-Thomé, Philipp, editors. Geological Survey of Finland, Special Paper 42: 2006. pp. 65-74.

[111] Römer H, Willroth P, Kaiser G, Vafeidis AT, Ludwig R, Sterr H, Revilla Diez J: Potential of remote sensing techniques for tsunami hazard and vulnerability analysis – a case study from Phang-Nga province, Thailand. *Natur Haz Earth Sys Sci.* 2012;12:2103-2126.

[112] Teodoro AC, Pais-Barbosa J, Veloso-Gomes F, Taveira-Pinto F: Evaluation of beach hydromorphological behaviour and classification using image classification techniques. *J Coastal Res.* 2009;2(56):1607-1611.

[113] Teodoro AC, Pais-Barbosa J, Gonçalves H, Veloso-Gomes F, Taveira-Pinto F: Identification of beach features/patterns through image classification techniques applied to remotely sensed data. *Int J Remote Sens.* 2011;32(22):7399-7422.

[114] Teodoro AC: Applicability of data mining algorithms in the identification of beach features/patterns on high-resolution satellite data. *J Appl Remote Sens.* 2015;9(1): 095095.

[115] Lyzenga DR, Malinas N P, Tanis F J: Multispectral bathymetry using a simple physically based algorithm. *IEEE Trans Geosci Remote Sens.* 2006;44(8):2251-2259.

[116] Kanno A, Koibuchi Y, Isobe M: Statistical combination of spatial interpolation and multispectral remote sensing for shallow water bathymetry. *IEEE Geosci Remote Sens Lett.* 2011;8(1)64-67.

[117] Teodoro AC, Almeida R, Gonçalves M: Independent Component Analysis (ICA) performance to bathymetric estimation using high resolution satellite data in an estuarine environment, in Remote Sensing for Agriculture, Ecosystems, and Hydrology XVI, Christopher M. U. Neale; Antonino Maltese, Editors, Proceedings of SPIE Vol. 9239 (SPIE, Bellingham, WA 2014), 923915.

[118] Su H, Liu H, Wang L, Filippi AM, Heyman WD, Beck RA: Geographically adaptive inversion model for improving bathymetric retrieval from satellite multispectral imagery. IEEE Trans Geosci Remote Sens. 2014;52(1): 465-476.

[119] Lesser MP, Mobley CD: Bathymetry, water optical properties, and benthic classification of coral reefs using hyperspectral remote sensing imagery. *Coral Reefs.* 2007;26(4):819-829.

[120] Su H, Liu H, Heyman W: Automated derivation of bathymetric information from multispectral satellite imagery using a nonlinear inversion model. *Mar Geodesy.* 2008;31(4):281-298.

[121] Lyzenga DR: Passive remote sensing techniques for mapping water depth and bottom features. *Appl Optics.* 1978;17:379-383.

[122] Lyzenga DR: Remote sensing of bottom reflectance and water attenuation parameters in shallow water using aircraft and Landsat data. Int J Remote Sens. 1981;2(1): 71-82.

[123] Lyzenga DR: Shallow-water bathymetry using combined lidar and passive multispectral scanner data. Int J Remote Sens. 1985;6(1):115-125.

[124] Minghelli-Roman A, Goreac A, Mathieu, S, Spigai M, Gouton P: Comparison of bathymetric estimation using different satellite images in coastal sea waters. Int J Remote Sens. 2009;30(21):5737-5750.

[125] Teodoro AC, Gonçalves H, Pais-Barbosa J: Bathymetric estimation through principal components analysis using IKONOS-2 data, in Remote Sensing for Agriculture, Ecosystems, and Hydrology XII, Christopher M. U. Neale; Antonino Maltese, Editors, Proceedings of SPIE Vol. 7824 (SPIE, Bellingham, WA 2010), 782419.

[126] Eugenio F, Marcello J, Javier Martin J: High-resolution maps of bathymetry and benthic habitats in shallow-water environments using multispectral remote sensing imagery. IEEE Trans Geosci Remote Sens. 2015;53(7):3539-3549.

3

Geo-spatial Technology for Landslide Hazard Zonation and Prediction

Dericks P. Shukla, Sharad Gupta, Chandra S. Dubey and Manoj Thakur

Additional information is available at the end of the chapter

Abstract

Similar to other geo hazards, landslides cannot be avoided in mountainous terrain. It is the most common natural hazard in the mountain regions and can result in enormous damage to both property and life every year. Better understanding of the hazard will help people to live in harmony with the pristine nature. Since India has 15% of its land area prone to landslides, preparation of landslide susceptibility zonation (LSZ) maps for these areas is of utmost importance. These susceptibility zonation maps will give the areas that are prone to landslides and the safe areas, which in-turn help the administrators for safer planning and future development activities. There are various methods for the preparation of LSZ maps such as based on Fuzzy logic, Artificial Neural Network, Discriminant Analysis, Direct Mapping, Regression Analysis, Neuro-Fuzzy approach and other techniques. These different approaches apply different rating system and the weights, which are area and factors dependent. Therefore, these weights and ratings play a vital role in the preparation of susceptibility maps using any of the approach. However, one technique that gives very high accuracy in certain might not be applicable to other parts of the world due to change in various factors, weights and ratings. Hence, only one method cannot be suggested to be applied in any other terrain. Therefore, an understanding of these approaches, factors and weights needs to be enhanced so that their execution in Geographic Information System (GIS) environment could give better results and yield actual ground like scenarios for landslide susceptibility mapping. Hence, the available and applicable approaches are discussed in this chapter along with detailed account of the literature survey in the areas of LSZ mapping. Also a case study of Garhwal area where Support Vector Machine (SVM) technique is used for preparing LSZ is also given. These LSZ maps will also be an important input for preparing the risk assessment of LSZ.

Keywords: Landslide, LSZ, Remote Sensing and Geographic Information System, Modeling, SVM, Garhwal Himalaya

1. Introduction

According to the information on the International Red Cross, there are roughly 200 major natural disasters that occur each year in the world. These natural disasters cause an annual average loss of nearly 130,000 persons, and more than 140 million normal lives are affected. The frequency of occurrences of these natural disasters has increased many times in the recent past, and its effects are becoming more severe in the coming years. The major attribute is being the population growth, urbanization/industrialization leading to climate change. In general, most of the "natural risks" are accentuated by humans themselves by direct or indirect interference with the nature. Understanding a natural disaster is very difficult as it is a very complex system that involves various controlling and contributing factors. This means that no easy, one-sided solutions can be found, but applying the holistic approach to tackle such problems could yield some beneficial results. Currently, many researches are being carried out to understand the phenomenon acting behind these natural disasters such as floods, tsunamis, cyclones, earthquakes, landslides, etc. So to combat these natural risks, the holistic concepts should be developed and applied, particularly to tackle landslide risk as landslides are one of the major environmental problems in our society.

The adverse impacts of climate change on developing countries have been highly consequential. High-magnitude flash floods and increased rains has been one of the pertinent causes of extensive landslides, which accounts for around 4.89% of the globally occurring natural disasters during the last two decades. The unplanned urbanization and development coupled with continued deforestation may be attributed to this rise in figure. Landslides are quite frequent along the tectonically active Himalayan region. In the year 1984, Varnes defined the term **hazard** as "the probability of occurrence of a potentially damaging phenomenon within a specified period of time and within a given area". When such spatial distributions of hazards are represented on maps into various classes, it gives zonation maps. Thus, landslide hazard zonation refers to the division of area into various classes, which is categorised on the basis of degrees of actual/potential hazard caused by landslides. Hence, hazard zonation forms a critical factor for effective landslide management and is used as a tool for planning mitigation measures. The preparation of the landslide hazard map requires the analysis of most determining factors that leads to soil failure. The preparation of landslide hazard zonation requires detailed landslide inventory, processes involved in slope instability, triggering factors and many other associated studies. Landslides may occur due to a variety of conditional and triggering factors such as change in slope angle, slope aspect, faults, lithology, deforestation, improper drainage system, rainfall, and earthquakes. Thus, this zonation can be carried out at various scales from national (1:1 million) to local (1:5000). Depending on the scale of map, the parameters/factors and their accuracy varies.

With the advent of satellite data and various sensors, the scope of remote sensing has increased widely. The bird's eye view of the area at moderate to fine resolution gives fast and quick information about the terrain. Clubbed with the spectral and temporal characteristics of the satellite, the ability to identify and recognise landslides for the preparation of inventory map has been improved a lot. Both visual as well as automatic processes are well developed for recognition of landslide features. This preparation of inventory map has been made more effective with recent developments of resolution merging where data from different sensors could be merged to obtain better, sharp and good resolution images. Not only in the identification of landslides but also in the preparation of other contributing and controlling factors, remote sensing plays a crucial role. The elevation data from DEM (Digital Elevation Model) are used for the preparation of slope, aspect, relief, curvature, etc., parameter that controls the behaviour of landslide as well as the slope stability/instability. Not only the optical and multispectral data but the Radar and SAR data are being used for the analysis of landslides. The interferometric SAR technique is capable of distinguishing very minute changes in elevation and slope; hence, it is used for the identification of higher-resolution and correspondingly smaller area. Data from various sensors, i.e. optical, multispectral, thermal and microwave/radar, are being used for landslide studies.

There are various methods for the preparation of Landslide Susceptibility Zonation (LSZ) such as based on Fuzzy logic, Artificial Neural Network, Discriminant Analysis, Direct Mapping, Regression Analysis, Neuro-Fuzzy approach and other techniques. These different approaches apply different rating system and the weights, which are area and factors dependent. Therefore, these weights and ratings play a vital role in the preparation of susceptibility maps using any of the approach. However, one technique that gives very high accuracy in certain might not be applicable to other parts of the world due to change in various factors, weights and ratings. Hence, only one method cannot be suggested to be applied in any other terrain. This chapter discusses the methods being used in the field of LSZ, what are the input parameters being used, what the accuracy is and how best the method map the LSZ. However, it should be kept in mind that most of these methods/analysis are based on landslide inventory of any area, so the first and foremost step for working towards LSZ should be preparation of landslide inventory. Finally, this chapter discusses a case study of application of geo-spatial technology for preparation of LSZ in Garhwal Himalayan region, which is tectonically very active and prone to landsliding.

2. Various Approaches for LSZ Mapping

2.1. Regression Analysis

People are normally interested in finding the relationship between different variables. For example, whether smoking causes lung cancer? Regression analysis is the statistical method of finding relationship between dependent/predicted variable (denoted as y) and independent/predictor variables (denoted as $x_1, x_2, \ldots, x_n$), where n denotes the number of predictor variables [1]. The true relationship between y and $x_1, x_2, \ldots, x_n$ can be approximated by the regression model as indicated in equation 1

$$y = f\left(x_1, x_2, \ldots, x_n\right) + \varepsilon \tag{1}$$

where ε is assumed to be a random error representing the discrepancy in the approximation. It accounts for the failure of the model to fit the data exactly [2]. Typically, regression analysis is used for one of these three purposes [3] viz. (i) Modelling the relationship between x and y, (ii) Prediction of target variable, and (iii) Testing of hypotheses.

There are three types of regression models:

Simple linear regression: It models the linear relationship between two variables; out of which one is dependent variable y, and other is independent variable x. In this model, regression equation is given as below in equation 2

$$y = ax + b + \varepsilon \qquad (2)$$

where a = slope of regression line, b = intercept and ε = random error. Simple linear regression is shown in Figure.1.

Multiple linear regression: There are many situations when result depends on one or more predictor variables. In such situations, simple linear regression is not sufficient to model the output, hence it requires a regression equation as given in eq 3, which models the linear relationship between one dependent variable y and more than one independent variables $x_1, x_2, \dots, x_n$. In this model, regression equation is given as below

$$y = a_1 x_1 + a_2 x_2 + \dots + a_n x_n + a_0 + \varepsilon \qquad (3)$$

where $a_1, a_2, \dots, a_n$ are regression coefficients, a_0 = intercept and ε = random error

After the determination of regression model, its parameters are estimated based on the collected data. This is called as parameter estimation and model fitting. Most commonly used method of estimation is called the least square method [1, 2, 3].

Nonlinear regression: When the relationship between dependent and independent variable cannot be modelled using straight line, nonlinear regression is used. For example, nonlinear regression model for growth of a particular organism (y) as a function of time (t) can be written as

$$y = \frac{\alpha}{1 + e^{\beta t}} + \varepsilon \qquad (4)$$

where α and β are model parameters and ε = random error. All nonlinear functions that can be transformed into linear functions are called linearizable functions [2, 3].

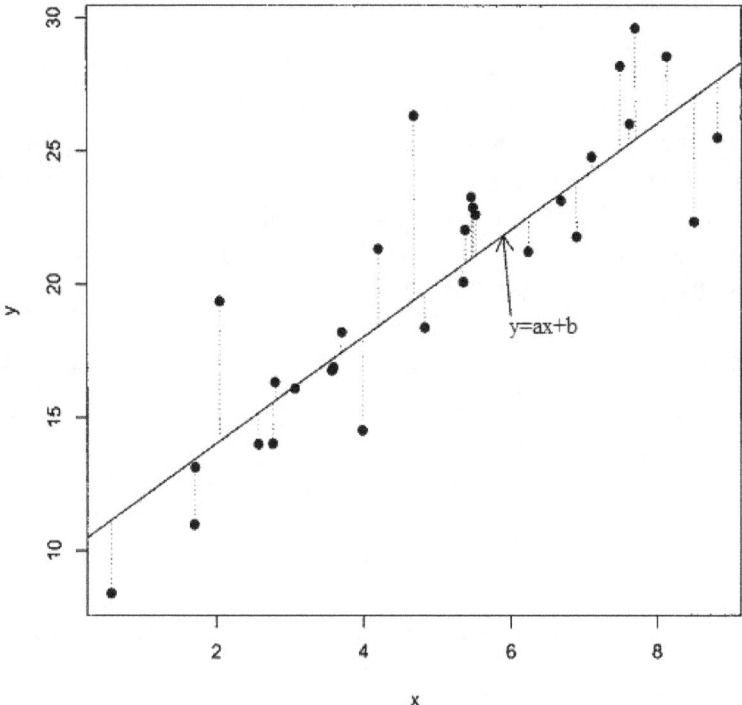

Figure 1. Simple linear regression model, solid line corresponds to true regression line and the dotted line corresponds to random error ε [3].

2.1.1. Estimation Using Least Square

The least square method for linear regression finds regression coefficients $a_0, a_1, a_2, ..., a_n$ such that sum of squared distance from actual value y_i and fitted value $\hat{y}_i$ reaches minimum for all possible choices of regression coefficients $a_0, a_1, a_2, ..., a_n$, [1, 4] using the given eq 5.

$$\sum_{i=1}^{n} \left[y_i - \left(a_0 + a_1 x_1 + a_2 x_2 + ... + a_i x_i \right) \right]^2 \tag{5}$$

For any choice of observed coefficients $\hat{a}$, the estimated/fitted value given for the observed values is

$$\widehat{y_i} = \widehat{a_0} + \widehat{a_1} x_1 + \widehat{a_2} x_2 + ... + \widehat{a_i} x_i \tag{6}$$

The difference between observed value y_i and fitted value $\hat{y}_i$ is called residual.

When dealing with regression analysis, if there is only one response variable, regression analysis is called univariate regression, and in case of two or more response variables, the regression is called multivariate regression. The difference between simple and multiple regressions is determined by the number of predictor variables (i.e. simple means one predictor variable and multiple means two or more predictor variables), whereas the difference between univariate and multivariate regressions is determined by the number of response variables. A brief summary of various classifications is given in Table-1. Out of all these regression types, logistic regression method is used a lot since most variables in hazard zonation mapping tends to be qualitative rather than quantitative.

Types of Regression	Conditions
Univariate	Only one quantitative response variable
Multivariate	Two or more quantitative response variables
Simple	Only one predictor variable
Multiple	Two or more predictor variables
Linear	All parameters enter the equation linearly, possibly after transformation of the data
Nonlinear	The relationship between the response and some of the predictors is nonlinear or some of the parameters appear nonlinearly, but no transformation is possible to make the parameters appear linearly
Analysis of Variance	All predictors are qualitative variables
Analysis of Covariance	Some predictors are quantitative variables and others are qualitative variables
Logistic	The response variable is qualitative

Table 1. Various Classifications of Regression Analysis [2].

2.1.2. Logistic Regression

Logistic regression model is a general linear model, which models the data with binary responses [1], i.e. it predicts the presence or absence of an outcome based on the values of a set of predictor variables [5]. The dependent variable in logistic regression is binary (i.e. 0 or 1, true or false), whereas the independent variable can be categorical, dichotomous or interval [6]. For landslide study, dependent variable is binary, showing either the presence or the absence of landslide.

Example: For determining risk factor for cancer, health data of several people were collected on several variables such as age, sex, smoking, diet, and the family's medical history. The response variable "y" is the person having cancer (y=1) or not having cancer (y=0) [2].

Coefficients of logistic regression can be used to calculate ratios for each independent variable in the model. Logistic regression model can be represented in simplest form as shown in equation 7

$$p = \frac{1}{1 + e^{-y}} \tag{7}$$

where p is the probability of occurrence of an event (varies between 0 and 1 on S-shaped curve), and y is dependent variable and calculated using the logistic regression equation 8

$$y = a_0 + a_1 x_1 + a_2 x_2 + \ldots + a_n x_n \tag{8}$$

where $a_1, a_2, \ldots, a_n$ are logistic regression coefficients and a_0 = intercept, $x_1, x_2, \ldots, x_n$ are independent variables [7].

2.1.3. Applications [2, 4]

i.   Agricultural sciences (e.g. analysis of data of milk production).

ii.   Management, industrial and labour relations (e.g. Do chief executive officers (CEOs) and their top managers always agree on the goals of the company?).

iii.   Environmental sciences (e.g. exploration of relationship between water quality and land use).

iv.   Psychology (e.g. What are the factors that impact the likelihood of a moonlighting worker becoming aggressive toward his or her supervisor?).

v.   Geography (Can the population of an urban area be estimated without taking a census?).

2.1.4. Landslide Hazard Zonation using Regression Analysis

Regression analysis is one of the most widely used statistical tool as it provides simple methods for establishing a functional relationship among variables. Logistic regression has been used widely for preparation of landslide hazard zonation maps [5, 6, 8, 9]. Slope, aspect, curvature, distance from drainage, lithology, distance from lineaments, land cover, vegetation index, and precipitation are considered as landslide-causing factors in many literatures. In logistic regression model, LHI is calculated by solving the regression equation. Correlation between landslide event and landslide affecting factors is estimated, and then, equation predicting the landslide is obtained.

2.2. Analytic Hierarchy Process

AHP, developed by Thomas L. Saaty in 1975, is an effective tool for decision making. It helps the decision makers in setting priorities and making best decision on complex decisive problems. It distributes the problems in hierarchy of criteria and options (alternatives), i.e. it reduces complex decisions to pairwise comparisons and then synthesizes the result. The AHP considers both the rational and the intuitive to select the best from a number of alternatives evaluated with respect to several criteria. It checks for consistencies in decision maker's evaluation and also allows limited inconsistencies in judgements.

2.2.1. Working of AHP

The AHP uses a set of evaluation criteria and a set of alternative options among which the best decision is to be made. It generates a weight for each evaluation criteria according to pairwise comparisons of criteria. The criteria with higher weight are selected since it is most important of all the criteria. Further, for fixed criteria, it assigns a score to each alternative option according to pairwise comparisons of options based on those criteria. Higher the score for an option, better the performance of that option w.r.t. considered criteria. Information is then arranged in a hierarchical tree. Finally, the AHP generates global score for each option using the combinations of the criteria weights and options scores and determines relative ranking of alternatives. A simple hierarchy with three levels is shown in Figure.2.

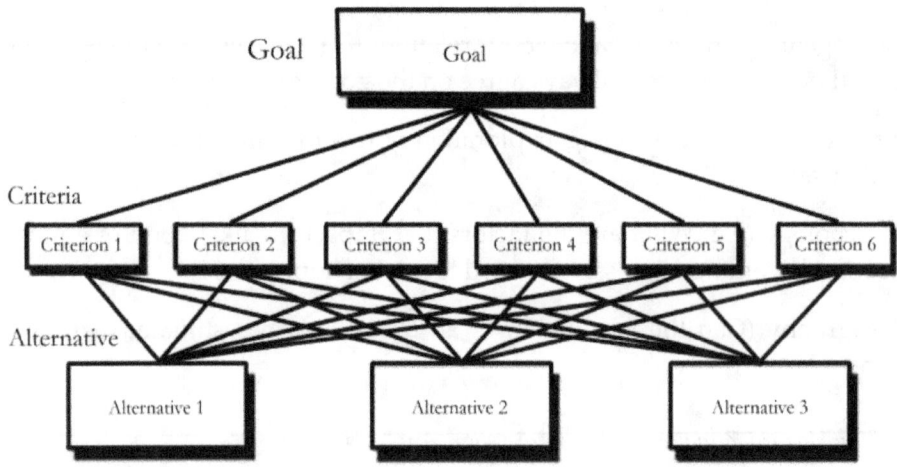

Figure 2. A three level hierarchy [10].

Implementation of AHP

AHP can be implemented in three simple steps

i.   Computation of weight vector for all criteria

ii.   Computation of score matrix for all options

iii.   Ranking of options based on final score

Once the goal has been set, then for all the alternatives, different ranks are given based on the criterion fixed to reach that goal. In this way, the priorities are set, and these factors are compared pairwise. For example, in case of landslide zonation, the goal could be to identify the areas that are prone to landsliding and the factors/parameters, such as slope, elevation, soil type, rock type, distance to drainage, etc., controlling it would become the alternatives. And to select the areas prone to landsliding, the criteria could be fixed such as slope should be more than 45°, soil type should be clayey, rock type should be other than granite/gneiss (hard rock), etc. Hence, the area fulfilling these criteria will be selected. This way of preparing the landslide susceptibility map is area specific, and the criteria applicable to one location may not be true for other location. Hence, a different approach is needed where the system adjust itself with the given conditions and scenarios.

2.2.2. The Fundamental Scale

The AHP is a general theory of measurement and is used to derive relative priorities of different criteria on absolute scales. Pairwise comparison judgments in the AHP are applied to pairs of homogeneous elements. The fundamental scale represents the intensities of judgments. In many cases, the elements to be compared are almost equal in measurements. In this situation, comparison must be made not on what fraction it is larger than the other [10]. Pairwise comparisons of criteria and/or options are performed based on the scale given in Table-2.

Intensity of Importance	Definition	Explanation
1	Equal importance	Two activities contribute equally to the objective
2	Weak	
3	Moderate importance	Experience and judgment slightly favour one activity over another
4	Moderate plus	
5	Strong importance	Experience and judgment strongly favours one activity over another
6	Strong plus	
7	Very strong or demonstrated importance	An activity is favoured very strongly over another; its dominance is demonstrated in practice
8	Very, very strong	
9	Extreme importance	The evidence favouring one activity over another is of the highest possible order of affirmation

Table 2. The fundamental scale by T. L. Saaty [10, 11].

2.2.3. Applications of AHP [10, 12, 13].

i.   Evaluation of cities for livelihood and planning

ii.   Ranking of countries

iii.   Customers adoption of mobile devices and mobile services

iv.   Human organ transplants

v.   Prediction of winners in chess matches

vi.   Natural resource management

2.2.4. Landslide Hazard Zonation using AHP

Various authors [14, 15, 16, 17, 18, 19] have used AHP for giving weights to various factors of landslide occurrence. The effect of each factor and factor classes, on landslide occurrence, is determined using pairwise comparison, and an equation is modelled for landside susceptible index (LSI), as given below in equation 9

$$LSI_{AHP} = \sum_{i=1}^{n} Factor_i * W_{AHP_i} \tag{9}$$

where $Factor_i$ = landslide conditioning factor such as slope, aspect, lithology, etc. W_{AHP_i} = Weightage for each causative factor. Pixel (LSI) values derived from above equation are classified into various susceptibility classes (low, moderate, high, and very high) based on natural break.

2.3. Artificial Neural Network

Artificial neural network attempts to model the information processing capabilities of the brain. The operation of the brain is based on simple basic elements called as neurons. Neurons are connected to each other with transmission lines called as axons and receptive lines called as dendrites. Information is stored at synapses. Each neuron has an activation level that ranges between some minimum and maximum value [20, 21]. A neural network is a massively parallel distributed processor made from simple processing units, which can store knowledge gained from experiments and can utilize it later. It replicates the processing of the brain in two respects [22].

i.   Knowledge is acquired by the network from its environment through a learning process.

ii.   Synaptic weights are used to store the acquired knowledge.

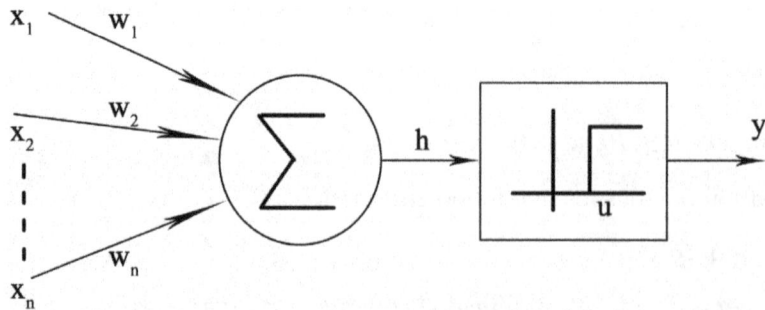

Figure 3. McCulloch and Pitts model of artificial neuron [20].

In 1943, McCulloch and Pitts proposed a computational model for artificial neuron, based on binary threshold [23]. This neuron calculates a weighted sum of 'n' input signals, x_j where j =

1, 2, 3.......n, and generates an output of 1 if this sum is above a certain threshold 'u', else output 0. The model [24] is shown in Figure. 3 and given by equation 10.

$$y = 1, if \ \sum_{j=1}^{n} w_j x_j > u,$$ (10)

$$y = 0, otherwise$$

ANN is a weighted directed graph, in which artificial neurons are nodes and directed edges with weights are connections between neuron outputs and neuron inputs. ANN can be grouped in two categories [20, 22].

i. Feed-forward network, where graph has no loops, as shown in Figure. 4. Here, all the nodes in each layers are connected to every other node in forward layer, hence it is called fully connected network. If some of the links are missing, then it is called partially connected network. Example: single-layer perceptron, multilayer perceptron, radial basis function, etc.

ii. Recurrent or feedback network, where graph has loops because of feedback connections, as shown in Figure. 5. Here, output from all the neurons is applied to input using feedback connection. Example: self-organizing map, adaptive resonance theory model, Hopfield network, etc.

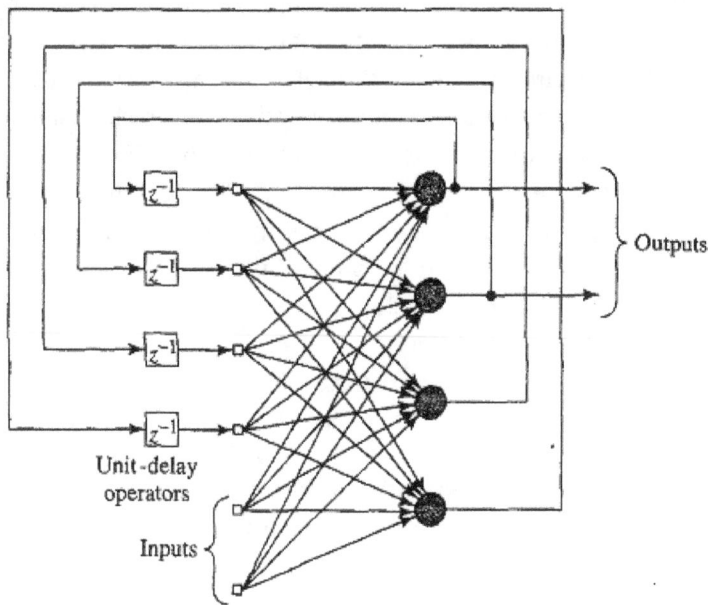

Figure 4. Example of feed forward network with one hidden layer & one output Layer

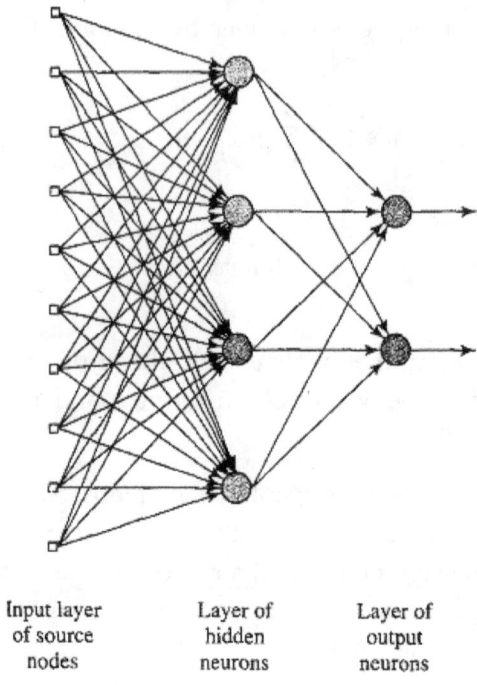

Input layer   Layer of   Layer of
of source   hidden   output
nodes   neurons   neurons

Figure 5. Example of recurrent network with hidden layer

2.3.1. Learning Algorithms

To be able to learn is the fundamental trait of intelligence. Although it is difficult to formulate a precise definition of learning, the process of learning in the context of ANN can be defined as the problem of updating network architecture and connection weights so that a network can efficiently perform a specific task [20]. Artificial neural network tries to learn input–output relationships from the given collection of representative examples, instead of following a set of rules specified by human experts. This is one of the major advantages of neural networks over traditional expert systems. A learning algorithm refers to a procedure in which learning rules are used for adjusting the weights. Some examples of learning algorithms are (i) Error correction learning, (ii) Memory-based learning, (iii) Hebbian learning, (iv) Competitive learning, (v) Boltzmann learning, etc. [23, 25].

2.3.2. Feed-Forward Back-Propagation Network (Based on error correction learning)

It is basically a feed-forward multilayer perceptron with back-propagation as learning/training algorithm. In order to train a neural network to perform desired task, the weight of each input has to be adjusted, such that the error between the desired and actual output is minimal (Figure. 6 after [21]) i.e.

$$\text{Error Signal}(e) = \text{Desired Response}(d) - \text{Actual Output}(y)$$

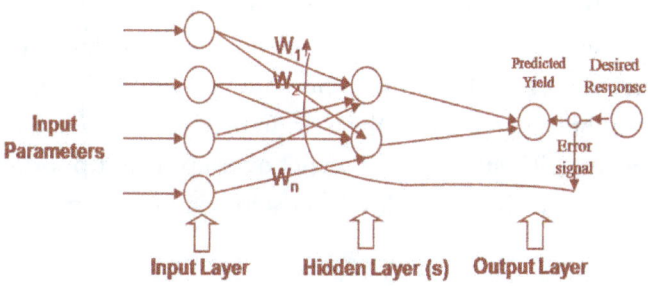

Figure 6. Back-Propagation Neural Network [21].

2.3.3. *Applications of ANN*

1. Image processing, classification of satellite data, compression of large images, etc.

2. Medical signal processing and developing medical decision support system [26, 27].

3. Paper making industry for prediction of curl in paper reel [28].

4. Calculation of nonlinear interpolation algorithm [29].

5. Detection and classification of vehicles in traffic management [30].

6. Optical and handwritten character recognition [31].

7. Operations research [32].

8. Application in Mineral Potential Mapping [33].

9. Landslide Susceptibility Mapping [34-37].

2.3.4. *Application of ANN in Landslide Hazard Zonation*

ANN has been used widely in the preparation of LHZ maps [34–37]. People have used variations of ANN with one input layer, two hidden layers, and one output layer for various factors controlling landslide occurrence. ANN connection weights are used to provide weights or rankings to the input data source (landslide-causative factors). Weights of factors and rankings of categories are integrated to provide LSZ map.

2.4. Support Vector Machine

Support Vector Machine is a data classification technique, developed by Vapnik in 1990. Classification process involves separating data into training and testing sets. Each element in the training set contains a corresponding target value (i.e. the class labels) and several attribute (i.e. the features of elements). The ultimate goal of SVM is to predict the target value for the test data, with only attributes of the test data given [38, 39]. Support vector machines are based on the concept of decision planes that define decision boundaries [40]. SVM finds the best hyperplane (n-dimensional plane) that separates all data points of one class from those of other class. It uses kernel method to project linearly non-separable data to a higher dimension. The kernel can separate classes even if mean values are near to each other. A simple illustration of the method is shown in Figure.7. The data points shown are linearly separable. The maximum margin hyper plane is shown in red, and the margin between the support vectors is shown by the parallel light blue lines. The two classes do not overlap. The support vectors (patterns that are on the margin) are shown [41] as yellow circles for class 1 and triangles for class 2.

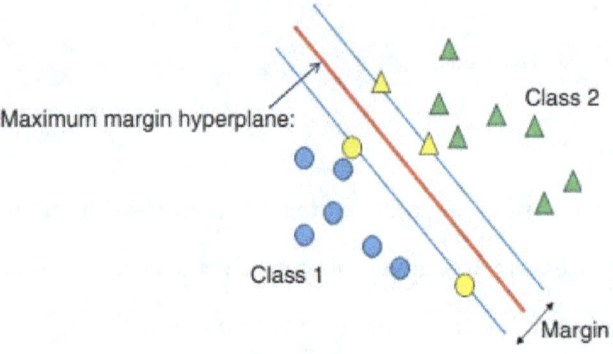

Figure 7. Illustration of the support vector [41].

Let m-dimensional training inputs x_i (i=1,...,M) belong to Class 1 or 2 and the associated labels be y_i = 1 for Class 1 and −1 for Class 2. If these data are linearly separable, we can determine the decision function, which is represented by equation 11 [42]

$$D(x) = w^T x + b \tag{11}$$

where **w** and b are weight and bias, respectively, to map the input into a higher dimensional space. The optimal separating hyper plane (i.e., $w^T x + b$ = 0) is located where the margin between the two classes is maximized, and the misclassification is minimized. The optimal hyper plane satisfies the following constrained minimization as given by equations 12–13

$$Min: \frac{1}{2} w^T w \tag{12}$$

$$w^T x_i + b \begin{cases} > 0 \ for \ y_i = 1, \\ < 0 \ for \ y_i = -1 \end{cases} \tag{13}$$

They can be obtained by solving the following constrained optimization problem by the method of Lagrange multipliers and maximizing the equation 14 as given below

$$L(w,b) = \frac{1}{2}(w.w) - \sum_{i=1}^{m}\alpha_i\left(y_i\left(w.x_i + b\right) - 1\right)$$ (14)

where α_i = Lagrange's multiplier and $\alpha_i \geq 0$. SVM can perform only binary classification; however, classifying data in more than two classes can be performed using pairwise classification [42, 43].

2.4.1. Applications of Support Vector Machine

i. Image processing for classification of satellite images [44].

ii. Modelling of Seismic Liquefaction Potential [45].

iii. Financial Literacy Modelling [46].

iv. Text charactization [47].

v. Face detection [48].

vi. Texture classification [49].

3. Advantages and Disadvantages

All these methods mentioned above have certain advantages as well as disadvantage over the other, hence a detailed comparative Table 3 showing their advantages and disadvantages are given below.

Method	Advantages	Disadvantages
Regression Analysis	Model developer has full knowledge of variables.	It requires the data to be independent.
	It is most strongly predictive of an outcome.	It is sensitive to outliers.
	It runs faster than neural network/support vector machine-based models.	
	It is not "black box" as ANN.	
Analytic Hierarchy Process (AHP)	It is simple, flexible and powerful.	It requires a large number of comparisons.
	All the calculations are driven by decision maker's experience.	Limitation of the use of 9 point T. L. Saaty's scale.
	It does not require an expert system with the decision maker's knowledge embedded in it.	It adds extra burden on decision maker for complex problem.
		Rank reversal

Method	Advantages	Disadvantages
Artificial Neural Network (ANN)	It requires less formal statistical training to develop the network.	Neural network are "black box".
	It can implicitly detect complex nonlinear relationships.	Single-layer perceptron work only on linearly separable classification problems.
	Availability of multiple training algorithms.	It requires greater computational resources.
		It is prone to over fitting.
		Can trap in local minima.
Support Vector Machine (SVM)	It has high prediction accuracy and good mathematical foundation.	The biggest limitation of the support vector approach is the choice of the kernel.
	Overfitting does not occur.	It requires long training time.
	It does not trap in local minima, i.e. it finds the global solution.	Problem has to be formulated as two-class problem.
	It works well with fewer training samples (i.e. number of support vectors do not matter much).	
	It requires fewer parameters (kernel, error cost).	

Table 3. Advantages and Disadvantages of these methods [10, 12, 13, 50, 51]

4. Literature Survey

The literature survey of some of the available research works carried out for Landslide Susceptibility Zonation is shown in Table 4 below:

S. No.	Techniques used	Accuracy (%)	References
1.	Discriminant Analysis	83.8	Carrara et al [52]
2.	Regression Analysis	70	Jade & Sarkar [53]
3.	Logistic Regression	74.8	Guzzetti et al. [54]
4.	Multilayer Perceptron	73	Ermini et al. [55]
5.	Neuro-Fuzzy approach	97	Pradhan et al. [36]
6.	Combined Neural Network and Fuzzy	74.5	Kanungo et al. [56]

Table 4. A comparative table for various techniques uesd with their accuracy.

The results obtained showed that the Artificial Neuro Fuzzy (ANF) modeling is a very useful and powerful tool for the regional landslide susceptibility risk assessments. Various membership functions should be selected and a number of training sets should be carefully and optimally selected to prevent over learning of the model. Therefore, the results that are to be obtained from the ANF modeling should be assessed carefully because the over learning may cause misleading results [35]. As a final recommendation, the results obtained from various papers showed that the methods followed in the study based on Neuro-Fuzzy approach exhibits a high performance. However, it is not forgotten that the performance of such type maps depends not only on the methodology followed but also on the quality of the available data and the factors considered for preparing LSZ. These input factors can be natural factors (like rainfall, lithology, slope, etc.) and anthropogenic factors (like road construction, mining, etc.). For this reason, if the quality of the data increases, the performance of the maps produced by these methods could increase. The detailed literature survey where various different models have been used for landslide hazard zonation is given below:

Lee and Pradhan[5] used frequency ratio and logistic regression model for mapping the landslide susceptible areas by considering slope, aspect, curvature, distance from drainage, lithology, distance from lineaments, land cover, vegetation index, and precipitation as landslide stimulating factors. They calculated the Landslide Hazard Index (LHI) by summation of frequency ratios for all the factors and solving the regression equation, respectively, for both methods and concluded that the frequency ratio model has 2.7% (93.04–90.34%) better predication accuracy than the logistic regression model.

Pradhan et al [57] combined frequency ratio and fuzzy algorithm for generating landslide hazard maps. Fuzzy membership values were calculated using frequency ratio and detected landslides. Fuzzy algebraic operators (such as fuzzy and, or, product, sum) and fuzzy gamma operators were applied on fuzzy membership values for landslide hazard mapping. Value of fuzzy gamma operator was set to 0.025, 0.05, 0.1, 0.2, 0.3, 0.4, 0.5, 0.6, 0.7, 0.8, 0.9, 0.95, and 0.975 for detecting its effect on landslide hazard maps. After verification, they found that out of 17 cases tested, the gamma operator with value 0.8 performed best (prediction accuracy 80.26%), while 'Fuzzy algebraic sum' and 'fuzzy or' showed worst accuracy of 64.77% and 56.86%, respectively.

Pourghasemi et al[14] showed the applicability of fuzzy logic and analytic hierarchy process in the mapping and zonation of landslide susceptible areas. A total of 12 data layers, which correspond to 12 landslide conditioning factors, were exploited to detect the most susceptible areas. Fuzzy membership values to all pixels were assigned based on the frequency ratio model. Landslide susceptibility was then identified using fuzzy if then else rules. Using the AHP model, weightage of each contributing factor was identified using pairwise comparisons and an equation was modelled for landside susceptible index. Validation of the maps created using both the methods was performed using ROC curve. They concluded that the model with fuzzy logic has the highest area under the curve (AUC) value 0.9194, whereas AHP has 0.8887.

Devkota et al[6] compared certainty factor, index of entropy and logistic regression methods for landslide susceptibility mapping. Slope gradient, slope aspect, altitude, plan curvature, lithology, land use, distance from faults, rivers and roads, topographic wetness index, stream power index and sediment transport index were considered as prominent factors for landslide susceptibility study. The value of the certainty factor ranges between −1 and +1. A positive value means an increasing certainty in landslide occurrence, while a negative value corresponds to a decreasing certainty in landslide occurrence. CF values of the landslide conditioning factors were combined pairwise to generate landslide susceptibility index. Natural breaks were used to classify LSI value to Landslide Hazard Zones. The performance of landslide susceptibility models was assessed using ROC curves. They found that the hazard map prepared using the index of the entropy model has the highest prediction accuracy (90.16%), followed by the logistic regression model (86.29%) and the certainty factor model (83.57%).

Nourani et al[8] prepared landslide hazard zonation maps using genetic programming and compared it with frequency ratio, logistic regression, artificial neural network. Seven factors, i.e. lithology, slope, aspect, elevation, land cover, distance to stream, and distance to road, were considered prominent for landslide hazard zonation study. In the frequency ratio model, landslide hazard index was calculated by summation of frequency ratios for all the factors. In the logistic regression model, LHI was calculated by solving the regression equation. Correlation between landslide event and landslide affecting factors was estimated, and then, equation predicting the landslide was obtained. Three layered feed-forward neural network with back-propagation as training algorithm was used for calculation of LHI. Two different criteria were used to measure the efficiency of the ANN method, i.e. the root mean square error (RMSE) and the determination coefficient (DC). For producing the best landslide susceptibility maps, sensitivity analysis was also implemented in ANN. For verification of LSM, produced by FR, LR, ANN, and GP methods, landslide testing data were compared with these maps. The assessment of AUCs showed that the prediction accuracy of FR, LR, ANN, and GP methods were 89.42%, 87.57%, 92.37%, and 93.27%, respectively.

Bui et al[37] compared the accuracy of landslide prediction, using support vector machine, multilayer perceptron neural network, radial basis function neural network, kernel logistic regression and logistic model tree. Slope, aspect, altitude, relief amplitude, topographic wetness index, stream power index, sediment transport index, lithology, fault density, land use, and rainfall were studied as landslide conditioning factors. For choosing the best subset of conditioning factors, predictive ability of the factors was assessed using the information gain ratio with 10-fold cross-validation technique. The analysis of landslide inventory map showed that landslides mainly occurred during and after the heavy rainfall. The performance of landslide susceptibility models was assessed using receiver operating characteristics (ROC) curves, and reliability was assessed using kappa index. They found that the MLP neural net model has the highest prediction capability of 90.2%, followed by the SVM model 88.7%, the KLR model 87.9%, the RBF neural net model 87.1%, and the LMT model 86.1%.

Youssef et al[9] combined logistic regression and frequency ratio for removing their weaknesses and producing landslide susceptibility maps with better accuracy. Altitude, curvature, distance from wadis, distance from road, distance from fault, stream power index, topographic wetness index, soil type, geology, slope, and aspect were used as contributing factors in landslide occurrences. Frequency ratio was calculated by analyzing the relationship between 11 conditioning factors and landslide occurrence. Landslide hazard index was calculated by summation of frequency ratios for all the factors and solving the regression equation, respectively, for the frequency ratio and logistic regression methods. After this, the probability index for ensemble of FR and LR was calculated and normalized to be between 0 and 1. For calcu-lating the landslide susceptibility map from ensemble method, the probability index value was classified in five categories using quantile classifier. Probability index value represents the predicted probability of landslide for each pixel in the presence of given set of conditioning factor. Validation of all three models was performed using ROC curves, and they observed that the prediction accuracy of ensemble of FR and LR was higher (82%) than that of FR (58%) and LR (77%) separately.

5. Case Study

The landslide susceptibility mapping is carried out in the Mandakini River basin of Uttarak-hand, which covers an area of about 2439 sq. km and is situated between 30°19'00"N to 30°49'00"N latitude and 78°49'00"E to 79°20'00"E longitude (Figure. 8a) falling in Survey of India toposheet Nos. 53J and 53N.

5.1. Geological setting of the Study Area

The lithological mapping of the area (Figure. 8b) shows the presence of Vaikrita formation in the north, forming most of the Greater/Higher Himalaya in Garhwal. South of this formation, the Munsiyari formation is present in the Lesser Himalaya. South of the Munsiyari formation, the Ramgarh group is present. The southernmost area of the basin is comprised of Beringay Formation. Vaikrita, Munsiyari, Ramgarh, and Beringay formations are, respectively, separated by Main Central Thrust (MCT-I), which is equivalent to Vaikrita Thrust; Main Central Thrust (MCT-II), which is equivalent to Munsiyari/Jutogh Thrust and Main Central Thrust (MCT-III), which is equivalent to Ramgarh/Chail Thrust [58, 59] (Figure. 8b). The presence of MCT Thrust zone causes high shearing and fractures in this area, which makes the rocks weak and highly prone to landslides and other natural hazards.

The high susceptibility to landslides in the Mandakini River basin is mainly due to complex geological settings, varying slopes and relief, heavy rainfall, along with ever-increasing human interference in the ecosystem. Extreme climatic events increase the instability of the terrain, which results in landslides, example includes the Kedarnath disaster [60]. Some of the major landslides occurred in the past are near Okhimath in 1997, 1998, 2010, 2012, 2013; in Phata Byung area in 2001, 2005, 2013; in Madhyamaheshwar area in 1998, 2005, 2013, etc., which are dependent on various factors such as geology, structure, land use, old slides, slope, slope aspect, and drainage in the area [61, 62, 63].

5.2. Data Used

The Survey of India (SOI) toposheet Nos. 53N and 53J were used to create the base map of the study area. Landsat satellite image of October 2008 with 30-m spatial resolution was taken to finalize the tectonic and geologic map of the study area (after) [59]. Elevation data were taken from ASTER-GDEM (Advance Spaceborne Thermal Emission and Reflection Radiometer, Global Digital Elevation Model) having spatial resolution of 30 m with an accuracy of ±10 m.

These data sets were analyzed, preprocessed and then categorized using Arc GIS 9.3, ERDAS Imagine 9.1 software to generate various thematic layers such as elevation, slope, aspect, drainages, geology/lithology, soil, buffer of thrusts/faults, and buffer of streams in the study area (Figure 8 a-h).

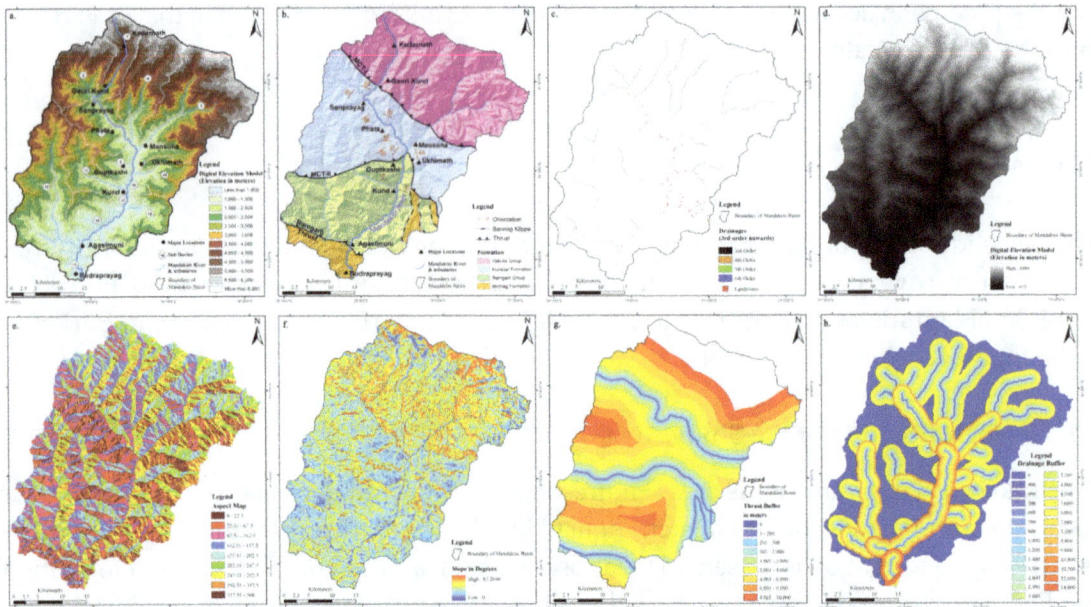

Figure 8. Various thematic layers used in landslide susceptibility prediction using PSVM model. a) Classified elevation map of the study area prepared from ASTER-GDEM showing major locations of Mandakini River basin. b) Geological map showing various formations and structures mainly MCT-I, MCT-II and Ramgarh Thrust (after Shukla et al. [59]). c) Drainage map derived from DEM showing third-order onwards and the presence of landslides in the study area. d) DEM map. e) Aspect map showing variation in the hill facets. f) Slope map showing comparatively higher slopes in northern sides as compared to southern side because of the presence of glacial features. g) Buffer map of the thrusts present in the study area created at specified intervals and reclassified in nine classes. h) Buffer map of the drainages third-order onwards. For the simplicity of the model, first- and second-order streams were not taken.

5.3. Model Selection and Results

All the data sets were generated in Geographic Information System (GIS) environment at 30 × 30 m pixel resolution, the vector layers were converted to raster format with other raster data sets. These raster data sets were converted to ASCII format to be read in MATLAB for using Support Vector Machine (SVM) for prediction of Landslide susceptibility. The landslide data for Okhimath River basin, procured from Geological Survey of India (GSI), were considered to test the SVM model and generate the predictive susceptibility map. The study area contains 1,805,548 pixels, while 2207 pixels are present as landslides. Thus, the pixels representing the landslides are mere 0.125% of the whole study area. The purpose of this study is to predict the landslide, so 1 denotes that pixel involved in landslide and −1 represents pixels that are not involved in landslide. In the whole study area, 2207 pixels were mapped as landslide based on the past data from GSI and other published reports. The whole set of data were divided into 60% as training data and 40% as testing data.

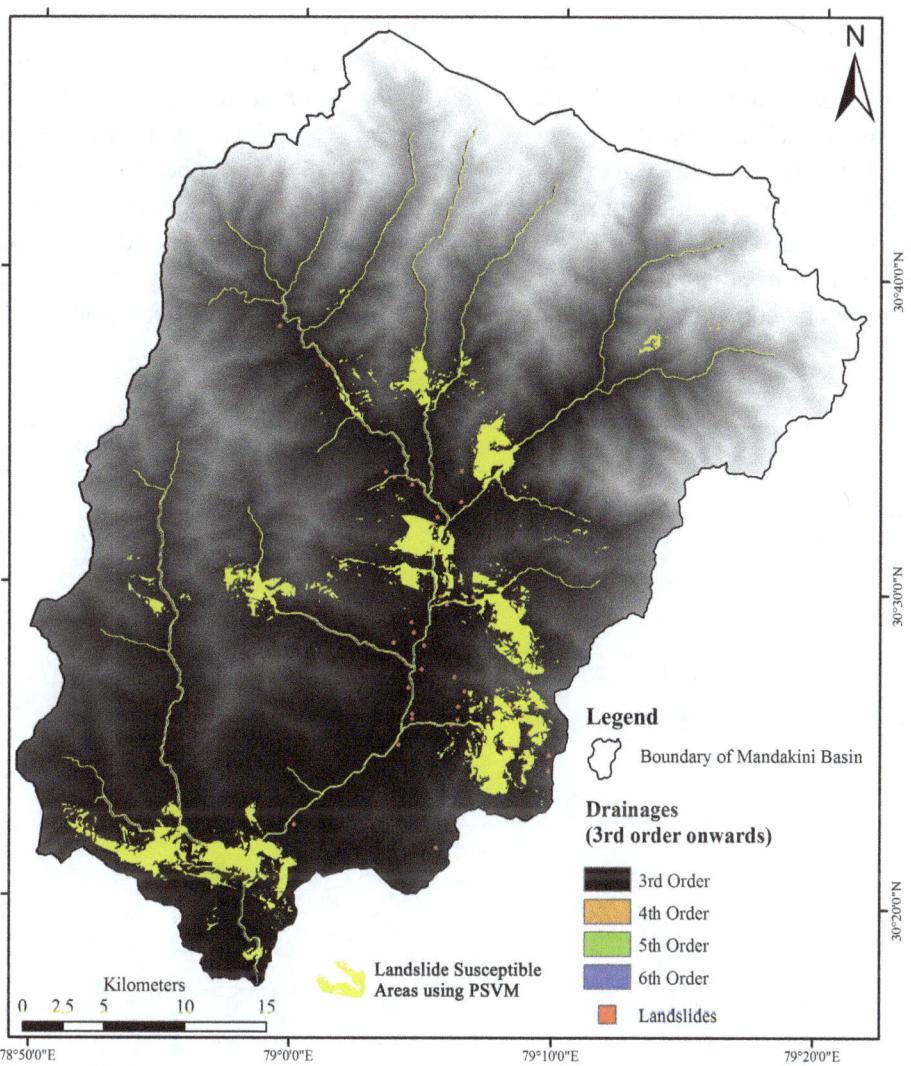

Figure 9. Landslide Susceptibility Map prepared using PSVM model shows areas susceptible to landslides on the DEM and drainage map of the study area with the actual past landslides.

Hence, the landslide susceptibility map for Mandakini River basin was prepared using the Proximal Support Vector Machine (PSVM) model (Figure. 9). It is evident from this figure that the PSVM model classified more areas in landslide susceptible zone as compared to certain landslides have been missed. Hence, various performance metrics such as average prediction accuracy (AA), true positive rate (TPR), true negative rate (TNR) and relative operating characteristic curve (ROC) were computed on testing data to validate the performance of prediction models [64, 65, 66]. The validation results in terms of AUC, and their corresponding testing accuracy showed that the PSVM model has higher AUC values when rainfall data from TRMM were considered with respect to when not considered as shown in Figure 10. The PSVM model with TRMM and without TRMM has an AA of 82.85% and 84.20%, TPR of 79.43% and 72.46%, TNR of 82.85% and 84.22% and an AUC value of 81.15% and 78.34%, respectively

(Table 5). The high value of TNR (82.85% and 84.22%) achieved by the PSVM model in this case is due to the large number of pixels for the study area as compared to pixels forming the landslides. Hence, this model predicted/demarcated the safe areas with 84.22% accuracy when TRMM data were taken into consideration, while it predicted the areas prone to landslide with 79.43% accuracy when TRMM data were taken in consideration because of less number of landslide pixels. Though the AUC values (78.34% and 81.15%) are good, the average accuracy for the PSVM model is quite high between 82.85% and 84.20%. Similar results were also obtained by Pradhan [67] where SVM yielded 81.46% AUC when applied on altitude, slope angle, plan curvature, distance from drainage, distance from road, soil type and NDVI as the input parameters considered for landslide susceptibility mapping for Penang Island in Malaysia.

Model	AA%	TPR%	TNR%	AUC%	C
PSVM (with TRMM)	82.85	**79.43**	82.85	**81.15**	100
PSVM (without TRMM)	**84.2**	72.46	**84.22**	78.34	128

Table 5. Prediction performance for PSVM model.

Best results are shown in bold. AA(%) is the average accuracy, TPR(%) is the true predictive rate, TNR(%) is the true negative rate and AUC(%) is the area under the curve.

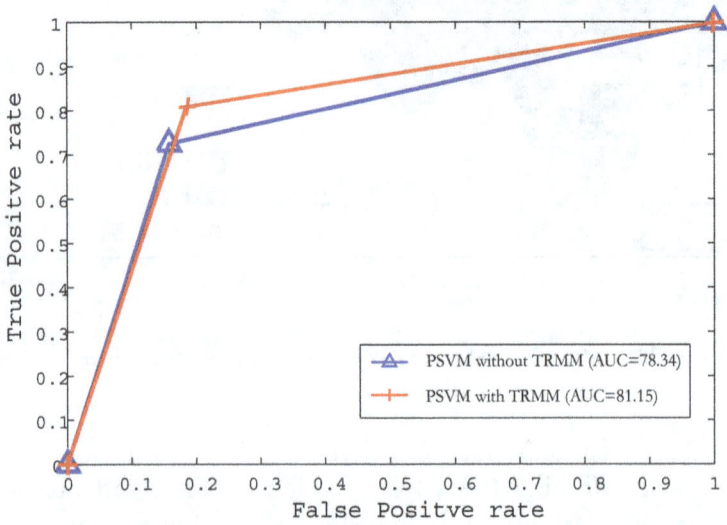

Figure 10. Best Prediction rate and area under the curves (AUC) produced by PSVM model with and without TRMM data consideration.

5.4. Conclusion

In Garhwal Himalaya, Mandakini River basin is highly vulnerable to landslides, especially the town of Okhimath and its nearby villages. In the vicinity of the study area, Mandakini

River crosses various Himalayan thrusts, and due to the presence of these tectonically active MCT zones, the rocks shows high shearing and fracturing and becomes more susceptible for landsliding. The susceptibility to landslide is mainly controlled by valley slopes, attitude of discontinuity of surfaces, soil type, presence of drainage, nature of rocks exposed, and structural and tectonic features present, besides human interaction in the terrain.

Hence, recently developed Support Vector Machine (SVM) learning technique was applied on this area to demarcate the landslide prone and safe areas. The PSVM method has been applied for landslide susceptibility mapping of the study area. The PSVM model showed higher average accuracy (AA) of 82.82%–84.20% for this study area, and the ROC curve indicates that the PSVM model has the prediction accuracy of 81.15%. Nevertheless, this model can be effectively used for landslide susceptibility mapping in this area or similar terrain with these sets of input parameters.

Acknowledgements

Authors would like to thank Dr. R. P. Singh, Ms. A. S. Ningreichon and Ms. Yogita Garbyal of Department of Geology, University of Delhi for carrying out the geological field mapping and figure preparations of this study area. The field work for this work was supported by DST project Landslide Dham (MANU Project), Project No. NRDMS/11/3010/013 (G) from NRDMS sanctioned to CSD.

Author details

Dericks P. Shukla[1*], Sharad Gupta[1], Chandra S. Dubey[2] and Manoj Thakur[3]

*Address all correspondence to: dericks.82@gmail.com

1 School of Engineering, Indian Institute of Technology, Mandi (HP), India

2 Department of Geology, University of Delhi, Delhi, India

3 School of Basic Sciences, Indian Institute of Technology, Mandi (HP), India

References

[1] Yan, X. & Su, X.G., 2009. Linear Regression Analysis: Theory and Computing, *World Scientific*, pp. 1–4.

[2] Chatterjee, S. & Hadi, A.S., 2006. Regression Analysis by Example. 4th ed., *Wiley Inter-Science*, New Jersey, pp. 12–15.

[3] Chatterjee, S. & Simonoff, J.S., 2013. Handbook of Regression Analysis. *Wiley Inter-Science*, New Jersey, pp. 3–16.

[4] Mendenhall, W. & Sincich, T., 2012. A Second Course in Statistics Regression Analysis. 7th ed., *Prentice Hall*.

[5] Lee, S. & Pradhan, B., 2007. Landslide hazard mapping at Selangor, Malaysia using frequency ratio and logistic regression models. *Landslides*, 4(1), pp. 33–41.

[6] Devkota, K.C., Regmi, A.D., Pourghasemi, H.R., Yoshida, K. et al., 2013. Landslide susceptibility mapping using certainty factor, index of entropy and logistic regression models in GIS and their comparison at Mugling-Narayanghat road section in Nepal Himalaya. *Natural Hazards*, 65(1), pp.135–165.

[7] Kleinbaum, D.G. & Klein, M., 2010. Logistic Regression: A Self-Learning Text, 3rd ed., *Springer*, pp. 4–10.

[8] Nourani, V., Pradhan, B., Ghaffari, H., & Sharifi, S.S., 2014. Landslide susceptibility mapping at Zonouz Plain, Iran using genetic programming and comparison with frequency ratio, logistic regression, artificial neural network models. *Natural hazards*, 71(1), pp. 523–547.

[9] Youssef, A.M., Pradhan, B., Jebur, M.N., & El-Harbi, H.M., 2014. Landslide susceptibility mapping using ensemble bivariate and multivariate statistical models in Fayfa area, Saudi Arabia. *Environmental Earth Sciences*, 73(7), pp. 3745–3761.

[10] Saaty, T.L. & Vargas, L.G., 2012. Models, Methods, Concepts & Applications of the Analytic Hierarchy Process. *Springer Science & Business Media*, New York, pp. 1–7.

[11] Saaty, T.L. & Kearns, K.P., 1985. Analytical Planning: The Organization of Systems. *Pergamon Press*, pp. 19–40.

[12] Saaty, T.L. & Vargas, L.G., 1982. The Logic of Priorities—Applications in Business, Energy, Health, and transportation. *Springer Science & Business Media*, New York.

[13] Brunelli, M., 2015. Introduction to the Analytic Hierarchy Process. *Springer Briefs in Operations Research*, pp. 1–15.

[14] Pourghasemi, H.R., Pradhan, B. & Gokceoglu, C., 2012. Application of fuzzy logic and analytical hierarchy process (AHP) to landslide susceptibility mapping at Haraz watershed, Iran. *Natural Hazards*, 63(2), pp. 965–996.

[15] Bhatt, P.B., Awasthi, K.D., Heyojoo, B.P., Silwal, T., & Kafle, G., 2013. Using geographic information system and analytical hierarchy process in landslide hazard zonation. *Applied Ecology and Environmental Sciences*, 1(2), pp. 14–22.

[16] Reza, M. & Daneshvar, M., 2014. Landslide susceptibility zonation using analytical hierarchy process and GIS for the Bojnurd region, northeast of Iran. *Landslides*, 11, pp. 1079–1091.

[17] Tazik, E., Jahantab, Z., Bakhtiari, M., Rezaei, A. & Alavipanah, S.K., 2014. Landslide susceptibility mapping by combining the three methods Fuzzy Logic, Frequency Ratio and Analytical Hierarchy Process in Dozain basin. In *ISPRS—International Archives of the Photogrammetry, Remote Sensing and Spatial Information Sciences*, pp. 267–272.

[18] Boroumandi, M., Khamehchiyan, M. & Nikoudel, M.R., 2015. Using of Analytic Hierarchy Process for Landslide Hazard Zonation in Zanjan Province, Iran. *Engineering Geology for Society and Territory*, 2, pp. 951–955.

[19] Arora, M.K., Das Gupta, A.S. & Gupta, R.P., 2004. An artificial neural network approach for landslide hazard zonation in the Bhagirathi (Ganga) Valley, Himalayas. *International Journal of Remote Sensing*, 25(3), pp. 559–572.

[20] Jain, A.K., Mao, J. & Mohiuddin, K.M., 1996. Artificial neural network: a tutorial. *Computer*, 29(3), pp. 31–44.

[21] Konar, A., 1999. Artificial Intelligence and Soft Computing, Behavioral and Cognitive Modelling of the Human Brain. *CRC Press*.

[22] Haykin, S., 2005. Neural Network: A Comprehensive Foundation. 2nd ed., *Prentice Hall*.

[23] Zurada, J.M., 1992. Introduction to Artificial Neural System. *West Publishing Company*.

[24] Demuth, H.B. & Beale, M., 2002. Neural Network Toolbox. *The MathWorks*, ver. 4.

[25] Alawala, C.R., 2007. Fuzzy Logic and Neural Networks: Basic Concepts and Applications. *New Age International Publisher*, pp. 121–143.

[26] Graupe, D., Liu, R.W. & Moschytz, G.S., 1988. Applications of neural networks to medical signal processing. In *Proceedings of the 27th IEEE Conference on Decision and Control Austin, Texas*, pp. 343–347.

[27] Gorzalczany, M.B., 1996. An idea of the application of fuzzy neural networks to medical decision support systems. In *Proceedings of the IEEE International Symposium on Industrial Electronics*, 1, pp. 398–403.

[28] Edwards, P.J., Murray, A.F., Papadopoulos, G., Wallace, A.R., et al., 1999. The application of neural networks to the papermaking industry. *IEEE Transactions on Neural Networks*, 10(6), pp. 1456–1464.

[29] Sun, Z., 2009. Application of neural network in calculation of nonlinear interpolation algorithm. In *IEEE International Conference on Information Science and Engineering*, pp. 3981–3984.

[30] Daigavane, P.M., Bajaj, P.R. & Daigavane, M.B., 2011. Vehicle Detection and Neural Network Application for Vehicle Classification. In *International Conference on Computational Intelligence and Communication Systems*.

[31] Mani, N. & Srinivasan, B., 1997. Application of artificial neural network model for optical character recognition. In *IEEE International Conference on Systems, Man, and Cybernetics. Computational Cybernetics and Simulation*. pp. 7–10.

[32] Smith, K.A. & Gupta, J.N.D., 2000. Neural networks in business: techniques and applications for the operations researcher. *Computers & Operations Research*, 27, pp. 1023–1044.

[33] *Lee, S. & Oh, H.J., 2011. Application of Artificial Neural Network for Mineral Potential Mapping, Artificial Neural Networks - Application, Dr. Chi Leung Patrick Hui (Ed.). ISBN: 978-953-307-188-6, *InTech*, DOI: 10.5772/16187.

[34] Kanungo, D.P., Arora, M.K., Sarkar, S., & Gupta, R.P., 2006. A comparative study of conventional, ANN black box, fuzzy and combined neural and fuzzy weighting procedures for landslide susceptibility zonation in Darjeeling Himalayas. *Engineering Geology*, 85(3-4), pp.347–366.

[35] Pradhan, B., Sezer, E. A., Gokceoglu, C., & Buchroithner, M. F. (2010). Landslide susceptibility mapping by neuro-fuzzy approach in a landslide-prone area (Cameron Highlands, Malaysia). *Geoscience and Remote Sensing, IEEE Transactions on*, 48(12), 4164-4177.

[36] Pradhan, B., Mansor, S. & Pirasteh, S., 2011. Landslide Susceptibility Mapping: an Assessment of the Use of an Advanced Neural Network Model with Five Different Training Strategies, Artificial Neural Networks - Application, Dr. Chi Leung Patrick Hui (Ed.), ISBN: 978-953-307-188-6, *InTech*, DOI: 10.5772/15738.

[37] Bui, D. T., Tuan, T. A., Klempe, H., Pradhan, B., & Revhaug, I. (2015). Spatial prediction models for shallow landslide hazards: a comparative assessment of the efficacy of support vector machines, artificial neural networks, kernel logistic regression, and logistic model tree. *Landslides*, 1-18.

[38] Hsu, C.W., Chang, C.C., & Lin, C.J., 2003. A Practical Guide to Support Vector Classification.

[39] Chang, C. C., & Lin, C. J. (2011). LIBSVM: A library for support vector machines. *ACM Transactions on Intelligent Systems and Technology (TIST)*,2(3), 27.

[40] Support Vector Machines (SVM) Introductory Overview, http://www.statsoft.com/Textbook/Support-Vector-Machines

[41] Mather, P.M. & Koch, M., 2011. Computer Processing of Remotely-Sensed Images: An Introduction. 4th ed., *Wiley Blackwell*, pp. 267-268.

[42] Abe, S., 2010. Support Vector Machines for Pattern Classification, 2nd ed., *Springer-Verlag* London, pp. 20-24.

[43] Campbell, C. & Ying, Y., 2011. Learning with Support Vector Machines. *Morgan Claypool Publishers*, pp. 1-5.

[44] Watanachaturaporn, P., Arora, M.K., & Varshney, P.K., 2008. Multisource Classification Using Support Vector Machines: An Empirical Comparison with Decision Tree and Neural Network Classifiers. *Photogrammetric Engineering & Remote Sensing*, 74(2), pp. 239–246.

[45] Samui, P., 2014. Vector machine techniques for modeling of seismic liquefaction data. *Ain Shams Engineering Journal*, 5, pp.355–360.

[46] Huang, R., Samy, M., Tawfik, H., & Nagar, A.K., 2008. Application of Support Vector Machines in Financial Literacy Modelling. *in Second UKSIM European Symposium on Computer Modeling and Simulation*, 2008, pp. 311–316.

[47] Moguerza, J.M. & Munoz, A., 2006. Support Vector Machines with Applications. Statistical Science, 21(3), pp.322–336.

[48] Osuna, E., Freund, R. & Girosi, F., 1997. Training Support Vector Machines: an Application to Face Detection. In *IEEE Conference on Computer Vision and Pattern Recognition*, pp. 130-136.

[49] Kim, K.I., Jung, K., Park, S.H., Kim, H.J., 2002. Support Vector Machines for Texture Classification. *IEEE Trans on Pattern Analysis and Machine Intelligence*, 24(11), pp.1542–1550.

[50] Tu, J. V, 1996. Advantages and disadvantages of using artificial neural networks versus logistic regression for predicting medical outcomes. *Journal of clinical epidemiology*, 49(11), pp.1225–1231.

[51] Igelnik, B., 2011. Computational Modeling and Simulation of Intellect: Current State and Future Perspectives. *Information Science Refer*, Chocolate Avenue Hershey PA, pp.226.

[52] Carrara, A., Cardinali, M., Detti, R., Guzzetti, F., Pasqui, V., & Reichenbach, P. (1991). GIS techniques and statistical models in evaluating landslide hazard.*Earth surface processes and landforms*, 16(5), 427-445.

[53] Jade, S., & Sarkar, S. (1993). Statistical models for slope instability classification. *Engineering Geology*, 36(1), 91-98.

[54] Guzzetti, F., Carrara, A., Cardinali, M., & Reichenbach, P. (1999). Landslide hazard evaluation: a review of current techniques and their application in a multi-scale study, Central Italy. *Geomorphology*, 31(1), 181-216.

[55] Ermini, L., Catani, F., & Casagli, N. (2005). Artificial neural networks applied to landslide susceptibility assessment. *Geomorphology*, 66(1), 327-343.

[56] Kanungo, D. P., Sarkar, S., & Sharma, S. (2011). Combining neural network with fuzzy, certainty factor and likelihood ratio concepts for spatial prediction of landslides. *Natural hazards, 59*(3), 1491-1512.

[57] Pradhan, B., Lee, S., & Buchroithner, M. F. (2009). Use of geospatial data and fuzzy algebraic operators to landslide-hazard mapping. *Applied Geomatics, 1*(1-2), 3-15.

[58] Ray, Y., Srivastava, P.: Widespread aggradation in the mountainous catchment of the Alaknanda-Ganga river system: timescales and implications to hinterland-foreland relationships. Quaternary Science Reviews 29(17), 2238-2260 (2010)

[59] Shukla, D., Dubey, C., Ningreichon, A., Singh, R., Mishra, B., Singh, S.: GIS-based morphotectonic studies of Alaknanda river basin: a precursor for hazard zonation. Natural hazards 71(3), 1433-1452 (2014)

[60] Dubey, C., Shukla, D., Ningreichon, A., Usham, A.: Orographic control of the Kedarnath disaster. Current Science 105(11), 1474-1476 (2013)

[61] Rautela, P., Thakur, V.: Landslide hazard zonation in Kaliganga and Madhyamaheshwar valleys of Garhwal Himalaya: a GIS based approach. Himalayan Geol 20:2, 31-44 (1999)

[62] Sati, S., Naithani, A., Rawat, G.: Landslides in the Garhwal Lesser Himalaya, UP, India. Environmentalist 18(3), 149-155 (1998)

[63] Chaudhary, S., Gupta, V., Sundriyal, Y.: Surface and sub-surface characterization of Byung landslide in Mandakini valley, Garhwal Himalaya. Himalayan Geology 31:2, 125-132 (2010)

[64] Bradley, A.P.: The use of the area under the roc curve in the evaluation of machine learning algorithms. Pattern recognition 30(7), 1145-1159 (1997)

[65] Brenning, A.: Spatial prediction models for landslide hazards: review, comparison and evaluation. Natural Hazards and Earth System Science 5(6), 853-862 (2005)

[66] Webb, A.R.: Statistical pattern recognition. John Wiley & Sons (2003)

[67] Pradhan, B.: A comparative study on the predictive ability of the decision tree, support vector machine and neuro-fuzzy models in landslide susceptibility mapping using GIS. Computers & Geosciences 51, 350-365 (2013)

4

Detection of Tree Crowns in Very High Spatial Resolution Images

Marilia Ferreira Gomes and Philippe Maillard

Additional information is available at the end of the chapter

Abstract

The requirements for advanced knowledge on forest resources have led researchers to develop efficient methods to provide detailed information about trees. Since 1999, orbital remote sensing has been providing very high resolution (VHR) image data. The new generation of satellite allows individual tree crowns to be visually identifiable. The increase in spatial resolution has also had a profound effect in image processing techniques and has motivated the development of new object-based procedures to extract information. Tree crown detection has become a major area of research in image analysis considering the complex nature of trees in an uncontrolled environment. This chapter is subdivided into two parts. Part I offers an overview of the state of the art in computer detection of individual tree crowns in VHR images. Part II presents a new hybrid approach developed by the authors that integrates geometrical-optical modeling (GOM), marked point processes (MPP), and template matching (TM) to individually detect tree crowns in VHR images. The method is presented for two different applications: isolated tree detection in an urban environment and automatic tree counting in orchards with an average performance rate of 82% for tree detection and above 90% for tree counting in orchards.

Keywords: Tree crown detection, VHR image, Template matching, Marked point process, Valley-following, Watershed segmentation, Local maxima, Region growing

1. Introduction

Inventories on forest communities are performed with the objective of providing support to the management and conservation activities in rural or urban forests or even in tree plantations. The traditional method of obtaining information on forest communities is to use systematic or random sampling or by sampling stands, so that the final parameters for the population are obtained on the basis of statistical extrapolation [1, 2]. Usually, the following parameters are determined for each tree included in the sampling: location, diameter at breast height (DBH), basal area (BA), height, identification of the species, crown size, and crown closure. Based on these measurements, other parameters such as volume of wood and biomass can be derived for the community stand. This renders the field survey techniques for forest inventories expensive, time consuming, and unsuited for large areas.

Remote sensing with high spatial resolution is a cost-effective and reliable way to obtain information about trees. It may be the only practical manner to assure sustainable management of forests with the necessary information, such as biochemical and biophysical data on the vegetation in a synoptic and repetitive manner for large areas and over long periods of time [3]. The tree crown is the basis of the data required for the inventory, for it allows to determine not only its size but also its position, crown closure, and, in some cases, the species. It also allows the derivation of parameters such as the density of the population, the health condition of the trees, the volume, the biomass, and the carbon sequestration rates [3–6]. This information is crucial to a series of applications such as the inventory and management of forested areas as well as in parks and urban forests. It can also be used for counting and monitoring trees in orchards or under power lines to prevent damage and accidents.

The study of individual trees with remote sensing started with the use of aerial photography with very high spatial resolution (scale greater than 1:10.000), driven mainly by the use of stereoscopy techniques. The task was performed by photointerpreters trained to recognize individual tree species, extract a series of measurements, or evaluate different types of damage [2]. The use of orbital optical remote sensing data for forest studies began in the 1970s, with the development of techniques to separate forested from non-forested areas [7]. The spatial resolution of these satellite images was the main limiting factor for more detailed studies about the forests, and as a result, the studies remained focused on the disturbances affecting forests (such as land clearing, burning, diseases, and pest) or to estimate some biophysical parameters of the vegetation [3,8]. It was only toward the end of the 1990s that orbital remote sensing began to provide very high resolution (VHR) data with a spatial resolution under 1 m, allowing the study of individual trees. Launched in 1999, Ikonos was the first of what is now a series of VHR satellites (Table 1), consolidating the use of orbital data for the study of individual trees. However, the increase in spatial resolution was not always accompanied by an increase in spectral resolution for VHR data which is often restricted to a single panchromatic band.

The increase in spatial resolution changed the focus of many remote sensing studies, which started to analyze not only classes of objects but also each object individually [9]. Branches and irregularities within the crowns became visible, and as a result, the spectral response of a tree is influenced by variations in the shape of the crown (differential illumination) and background effects. This causes an increase in the intra-class variance and often results in a reduced accuracy when using conventional pixel-based classification [10]. This had a significant effect on the image processing techniques for forest studies and generated the development of new forms of information extraction.

Within the study of individual objects, the automatic detection and delineation of tree crowns using remote sensing VHR imagery have attracted much attention from researchers in forestry and computer vision [4,7]. Researchers have developed several automatic and semiautomatic methods for extracting individual trees and their characteristics using digital aerial photos of various types and VHR satellite images [11]. The applications range from the identification of tree crowns to their delineation and classification and are often based on image segmentation algorithms and other advanced image processing and analysis techniques [9,12]. Most of these algorithms were specifically developed for the detection and delineation of trees in temperate forests based on the assumption that the trees are cone shaped and round (mostly coniferous) in the images, with the apex of the tree having the highest reflectance of the crown area [4].

Satellite	Launch Year	Px resolution* (m)	Mx resolution* (m)	Multispectral bands
Ikonos II	1999	0.82	3.2	Blue, Green, Red, Near IR (4)
QuickBird	2001	0.65	2.62	Blue, Green, Red, Near IR (4)
WorldView-1	2007	0.46	-	-
Geoeye-1	2008	0.46	1.84	Blue, Green, Red, Near IR (4)
WorldView-2	2009	0.46	1.85	Coastal, Blue, Green, Yellow, Red, Red Edge, Near IR, Near IR2 (8)
Pleiades 1A	2011	0.5	2.0	Blue, Green, Red, Near IR (4)
Pleiades 1B	2012	0.5	2.0	Blue, Green, Red, Near IR (4)
Kompsat-3	2012	0.7	2.8	Blue, Green, Red, Near IR (4)
SkySat-1	2013	0.9	2.0	Blue, Green, Red, Near IR (4)
WorldView-3	2014	0.31	1.24	Coastal, Blue, Green, Yellow, Red, Red Edge, Near IR, Near IR2 (8)
SkySat-2	2014	0.9	2.0	Blue, Green, Red, Near IR (4)
Kompsat-3A	2015	0.55	2.2	Blue, Green, Red, Near IR (4)
WorldView-4	2016	0.34	1.36	Not available at time of printing

* Panchromatic (Px) and Multispectral (Mx) resolution at nadir.

Table 1. Very High Resolution satellites (1999-2006), with their spatial resolutions and spectral bands. Note that there is not any available satellite with VHR in multispectral bands.

The analysis of individual trees based on remote sensing images is a complex problem. Images of trees with varied crown size increase the difficulty of the analysis. What is detected as a single object may in fact represent a separate branch or even a group of trees [10]. Other sources of error are caused by the proximity between neighboring trees, trees located under other trees, trees in the shade, or trees that have a low spectral contrast with the background [13]. Consequently, high-level complex algorithms are necessary to exploit this contextual information [1].

This chapter provides an overview of the state of the art in individual tree crown detection based on optical VHR remote sensing data. An original method developed by the authors is also presented as an alternative approach to the problem of tree crown detection. In Part I, we present the main algorithms developed for the detection of individual trees, be it for tree identification or delineation. The principle of each approach is presented as well as its potential and limitations. Part II is dedicated to outlining the original MPP–TM approach, a hybrid method that combined two methods used in pattern recognition: marked point process and template matching. The results are shown for tree detection and delineation in an urban environment and for tree counting in orchards.

2. Part I – Review of tree crown detection methods

We present six of the main algorithms used in individual tree detection in high spatial resolution images. The algorithms are summarily described individually, but it should be noted that many approaches use hybrid methods for the detection and delineation of tree crowns. For instance, some authors might use one algorithm for detecting the trees and another to delineate them; some may even use one approach as a first approximation and another to fine-tune the results.

2.1. Local maxima filtering

Local maxima (LM) filtering is a technique used for identifying tree crowns in high spatial resolution imagery which is based on the recognition of the points with the greatest brightness within a search window that scans the entire image [4,14]. The search window, with a fixed size, defines which pixel has the greatest reflectance compared to all the other pixels inside the window. The pixels with the highest digital number are identified as possible tree locations. This method is adequate for trees which have the greatest reflectance at their top, surrounded by lower intensity pixels, and due to its concept, it is widely used for detecting conifers.

When the kernel window passes over the image, it does not take into account the presence of trees with different crown sizes, and the success of the LM tree recognition depends on the careful selection of the size of the search window. If it is too small, errors of commission occur by selecting nonexistent trees or multiple radiance peaks for an individual tree crown; if it is too big, the algorithm is likely to miss some trees (omission errors) [13].

The identification of trees by LM is affected by false bright pixels, which are not part of the brightest part of the crown. An effective method for dealing with the problem is to apply a Gaussian filter to the image. This allows the low-pass filter function to grant more weight to the crown center pixels (surrounded by much lower values) compared to those located toward the crown edge which might belong to other bright objects or noise. Applying a Gaussian filter directly affects the number of local maxima identified and causes the smoothing off of the brightness values on the tree crown edges [15].

In order to minimize the problem of the window size with LM, reference [13] used windows of varying sizes based on the assessment of the spatial structure of the image obtained by analyzing the local semi-variogram with different pixel lags and different window sizes. This results in a personalized window for each pixel, leading to greater accuracy when compared to using a single fixed window size. Reference [16] used LM to identify the centroid of eucalyptus trees in Australia. The search for the trees is carried out based on the maxima in four linear kernels pertaining to the four main directions (0°, 45°, 90°, and 135°) of the image and by summing the individual maxima found in each pass (Figure 1).

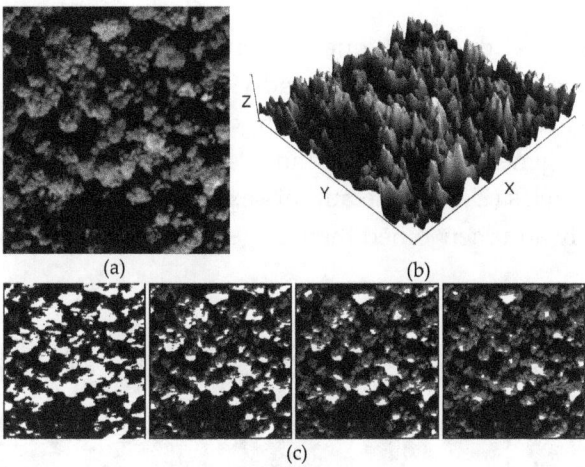

Figure 1. Examples of the surface produced by applying a LM kernel operator: (a) original image, (b) the local maxima appearing in the third dimension are associated with the presence of trees, (c) application of local maxima filter in four linear cumulative kernels (0°, 45°, 90°, and 135°).

2.2. Template matching

Template matching (TM) is a technique used for object recognition widely cited in the specialized literature which uses quantitative descriptors, such as length, area, and texture to describe recurring patterns in an image [17,18]. Based on a synthetic model or a sample extracted from the image, the correlation coefficient between the model and the image is calculated in order to determine the strength of the match between the two matrices. The object is assumed to be located where the measurement of the match reaches a maximum [4].

For tree crown detection, the study of reference [19] was the first to propose an elliptical 3D model for tree crowns based on an ellipsoid of generalized revolution (Equation 1).

$$\frac{z^n}{a^n} = \frac{\left(x^2 + y^2\right)^{n/2}}{b^n} = 1 \tag{1}$$

where z is the vertical axis of the center of the tree crown in its origin, a is half the height of the ellipsoid, b is half the radius, and n is the parameter of the shape of the tree crown. Subsequently, the model is illuminated using the acquisition parameters of the image (sun elevation and azimuth) and the characteristics of crown absorption and reflection of light in the chosen spectral band.

Because it is based on a physical model (rather than a complex mathematical concept), TM is considered a user-friendly method. Its limitation is mainly due to the need to use a library of models if many types of trees are present in the image, which may involve a complex phase for generating the models. Figure 2 shows examples of synthetic tree models and an application in a orchard.

References [20 and 21] used this technique to identify tree crowns in aerial images. Other researchers used this technique to recognize individual tree crowns, using templates made from small sub-images of the actual scene to identify the trees [22, 23]. Reference [24] proposed an improved version by generating separate models for trees and their shade in VHR images of unmanned aerial vehicles (UAV). The authors explored the relation between the tree and shade models separately and then joined them to generate a more powerful object detector.

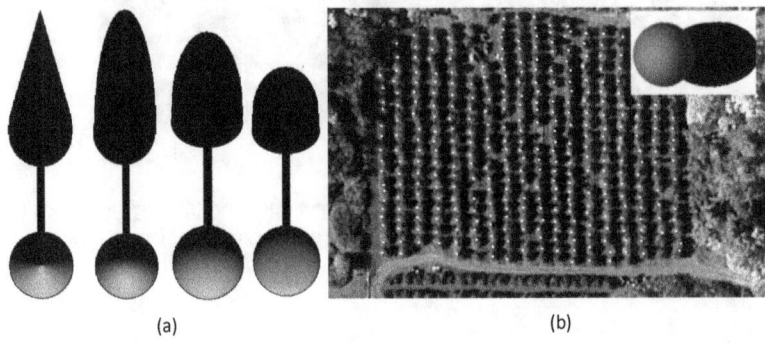

(a)   (b)

Figure 2. Left: examples of synthetic tree models to different tree crown shape. Right: identification of trees in an apple orchard showing the model used at the upper right corner.

2.3. Valley-following

Valley-following (VF) is a crown delineation method which identifies the shaded areas between the trees. This methodology was initially described in reference [25] and makes an analogy with topographic data, where the shades of gray of the pixels represent local lows in the third dimension. In this analogy, the bright tree crowns would be the hills and the darker zones around the trees the valleys (Figure 3). This darker zone is the one which typically helps human interpreters to separate one tree crown from the other. In this approach the shaded areas are eliminated, making it possible to separate the trees in the image. This was not sufficient to separate all of the trees, so the authors developed an approach based on a series of rules (e.g., no discontinuity, checking directions, context, gap filling, etc.) to accurately describe the boundaries of each tree, one at a time [26].

This approach performed well in images with a combination of low solar elevation angle and conical trees. Conversely, the approach failed to produce good results when the canopy was composed of trees of very different sizes, or when the tree crowns were very large and have internal shadows. The latter case resulted in subdividing the individual tree into two or more parts. Smaller trees, in contrast, tended to be grouped together. Reference [27] found that this

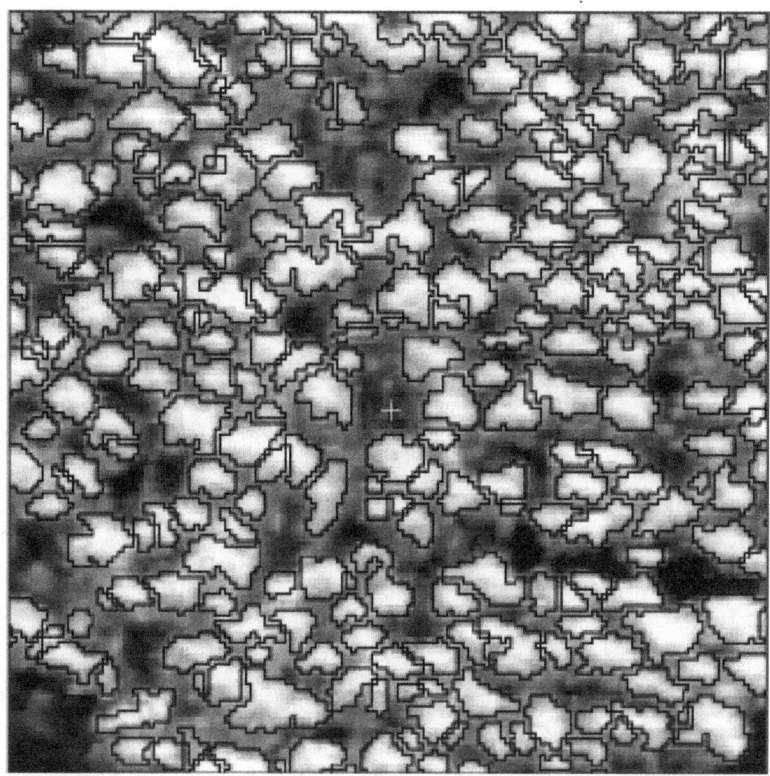

Figure 3. Results of the valley-following applied on a forest image in Canada (source: Gougeon and Leckie, 2003; reproduced with permission from Natural Resources Canada).

approach causes many false positives (FPs) in open areas (clearings). As a solution, they suggested the exclusion of these areas by retaining only the high-value pixels in the normalized difference vegetation index (NDVI).

2.4. Watershed

Like VF, the watershed segmentation (WS) is a technique related to thresholding that uses the gray levels in the images as if it were a topographic surface [28]. It is used not only for the delineation of individual tree crowns but also for generic segmentation of images. The watershed concept is based on a 3D image representation, with the third dimension being provided by the intensity of gray. The main objective of the watershed algorithm is to find the "drainage" divide lines. The "relief" in the image is inverted (high gray values become valleys) and progressively filled with a virtual liquid, and when the liquid is almost overflowing from one basin to another, a virtual dam is built, to create the watershed. These lines are considered the limits of each segment. The simplest approach to the construction of the dam is the use of morphological dilation of the minima, without merging the regions [17].

The images are usually preprocessed before the WS is applied. In fact, this segmentation is frequently applied to the gradient of an image, and not to the image itself. This is due to the relative homogeneity of the gray values of objects that do not provide sufficient contrast for an effective segmentation. In this formulation, the regional minimum value of the catchment basins usually correlates well with the lower gradient values that match the contours of the objects of interest [17,28]. The direct application of the WS algorithm generally leads to over-segmentation due to noise or other local irregularities of the gradient (Figure 4a). One of the approaches used to limit the number of regions is to use markers. The selection of markers can be based on simple procedures, intensity and connectivity between pixels, or even complex descriptors, such as size, format, location, relative distances, texture, and others. The use of markers provides prior knowledge to support the segmentation process [17].

The approaches that use WS for the delineation of the tree crowns normally use markers representing the center of the tree crown, to assist the segmentation process. For instance, reference [29] used WS to detect and delineate tree crowns in a VHR forest image in Canada but divided the approach into two phases, namely using LM to detect the crown and applying WS for the delineation. The LM image with the detected crowns was produced by using a Laplacian of Gaussian edge detection operator. The tree crowns were modeled based on their geometry and radiometry, resulting in an image of markers. This image then served to guide the WS in delineating the crowns. Reference [30] developed a bitemporal procedure for the automatic segmentation and reconciliation of groups of pixels (called blobs) within the forest using WS. By using two dates, they were able to increase the probability of properly defining the tree contours. Many problems were encountered in the segmentation process of the individual trees. For instance, trees with spread branches were sometimes split into two or more segments or contrarily by including several crowns in the same segment when trees were not sufficiently separated.

2.5. Region growing

Region growing (RG) is another segmentation technique that groups pixels or groups of pixels based on predefined growth criteria in an attempt to separate and recognize objects in the image [4]. Like WS, RG is used as a generic segmentation method and can be adapted for the delineation of individual tree crowns (Figure 4b). Starting with some seed pixels (which can be random if no other information is provided), the neighboring pixels are examined one by one and added to the growth region if their predefined properties are similar to those of the seeds (such as specific intervals of intensity or color) [17]. When no more pixels can be added or some predefined limit is reached (e.g., number of pixels), these pixels are labeled as belonging to the specific region of the seed pixel. Additional criteria can increase the power of an RG algorithm by introducing a higher concept like size and similarity between candidate pixels and the pixels selected or even the format of the region [17,31].

Reference [16] used RG integrated with LM to identify and delineate tree crowns in Australia. The LM method served to find the center of potential trees, which were then used as seeds for the RG. Reference [6] tested two different types of segmentation by RG, one by Brownian motion and the other by random walk, to detect conifers in a boreal forest. The methods were capable of detecting about 80% of the illuminated portion of the crowns, with a better per-formance found in larger crowns (Figure 5).

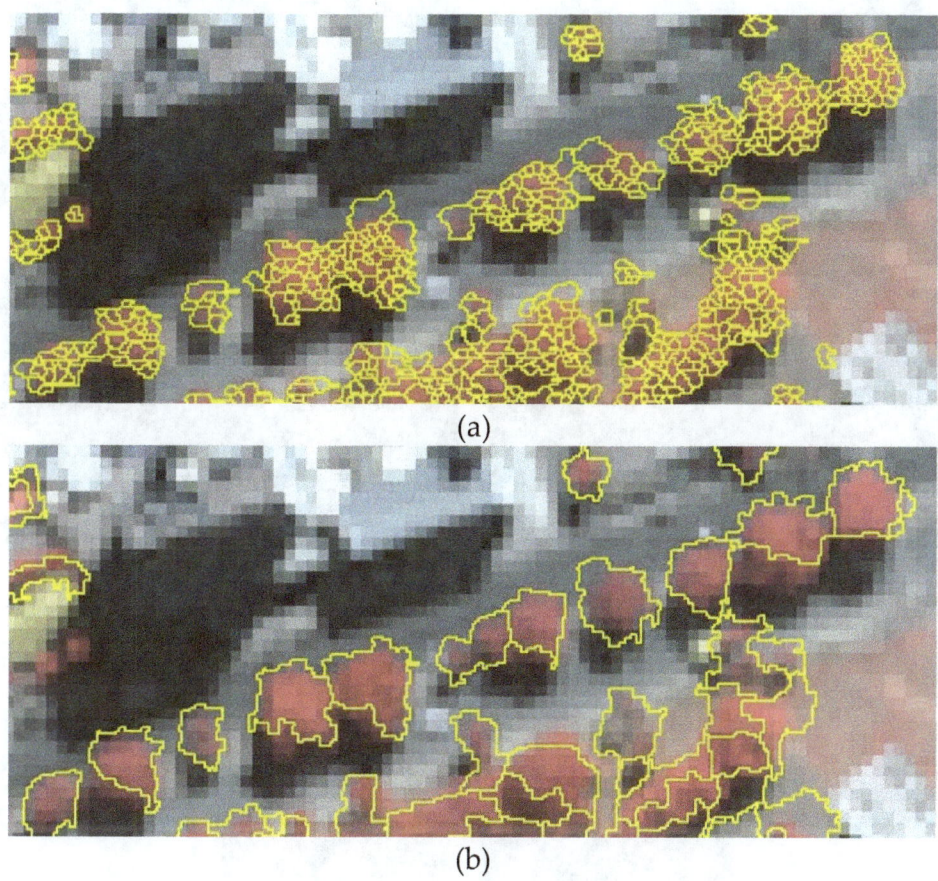

(a)

(b)

Figure 4. Comparison between two segmentation algorithms: (a) watershed and (b) region growing on a WorldView-2 image (panchromatic band with a 50 cm resolution). The WS was applied to the gradient image without using markers and resulted in the over-segmentation of the tree crowns. In (b), the RG segmentation was performed within an object-oriented classification (GEOBIA) approach, where the correct delineation of the tree crowns is noteworthy (source: Gomes and Maillard, 2013).

2.6. Marked point processes

The marked point process refers to a probabilistic method which has been used in recent years for the recognition of objects in high spatial resolution imagery [5,11,32–35]. In an MPP, sets of random points in a given space (x, y) are provided with a mark which is complete and separable, allowing the definition of a topology (defined by the mark) and the attribution of a label. An image is considered a random model where the gray tones are the realization of a random point process [34]. This random configuration of gray levels in the images is then modeled based on geometric figures (ellipses, circles, rectangles, and lines), respecting certain geometric (nature of the objects) and radiometric constraints (type of image).

The laws of density and probability distinguish various types of point processes, which can be Poisson, Strauss, Markov, or Gibbs, among others. The Markov or Gibbs point processes have been used for the recognition of tree crowns by a number of authors [5,32,33]. These

Figure 5. Example of RG segmentation to delineate tree crowns in a boreal forest. (a) The original image (with a spatial resolution of 3 cm), (b) with the results using Brownian motion, (c) and random walk (Source: Erikson, 2004).

processes are defined by a density function using a form of energy expressed as a sum of the a priori energy and the local energy. The process seeks to minimize the global energy of the model, by iterating it with some optimization scheme (Markov random fields, algorithm of multiple births and deaths, and Monte Carlo-Monte Carlo simulations).

Reference [5] proposed two different models to serve as marks in an MPP, one in 2D for detection of trees in densely forested zones (Figure 6) and the other in 3D for scattered or isolated zones, based on aerial photos of high spatial resolution in the infrared band. The MPP was integrated with a reversible jump Markov Chain Monte Carlo in a simulated annealing

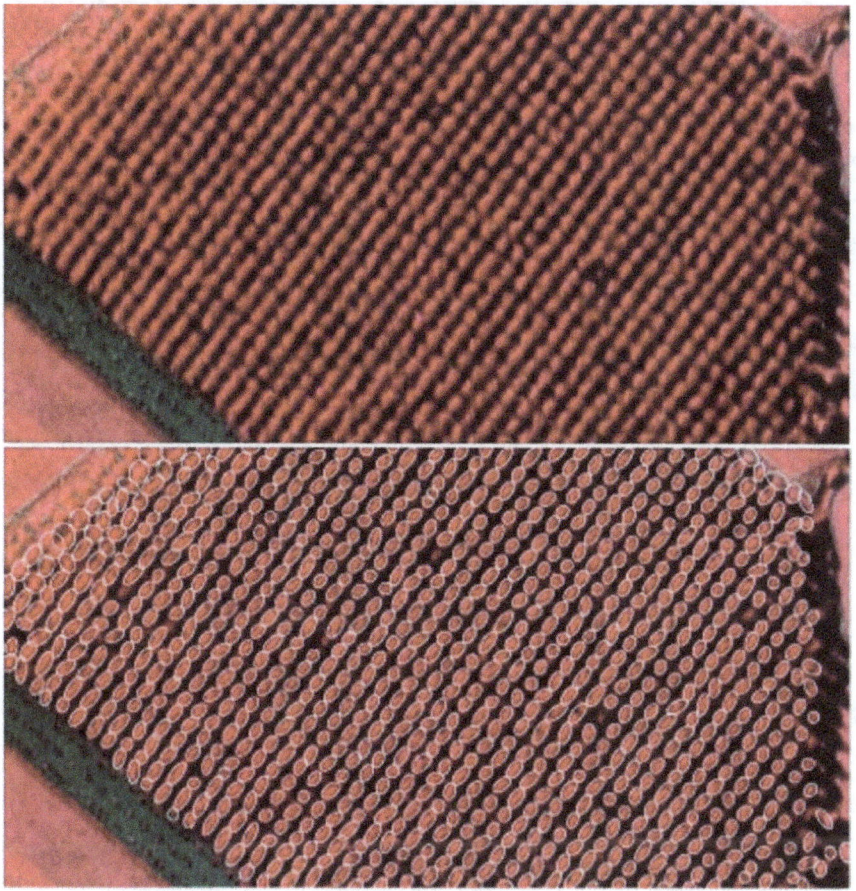

Figure 6. Example of the application of MPP for tree crown recognition on a poplar plantation. The original image is shown at the top and the results at the bottom (Source: Perrin, 2006).

method. Reference [32] used an MPP to automatically detect the tree crowns in high spatial resolution images, based on the modeling of the crowns as 2D circles on high-resolution spatial images. The method was successfully tested on mangrove forests and eucalyptus plantations.

2.7. Discussion

In the previous section, we have presented some of the most common algorithms used in the detection of individual trees, be it for their identification, delineation, or both. Table 2 presents a summary of these principles through their main characteristics and limitations.

Trees may differ in shape, size, spectral properties, height, foliage type, and density, and their spatial context varies with illumination, ground type, and inclination. They can also be surrounded by many other objects, especially in an urban setting. As such, the task is not trivial and can become highly complex depending on the number of parameters involved. Conversely, in planted forest and orchards where trees have the same age and species, tree crown extraction can take advantage of their relative uniformity.

Algorithm	Usage	Principle	Researches	Characteristics / Limitations
Local Maxima (LM)	Identification of tree crown	Identification of brightest points locally as corresponding to the apex of a treetop within a search window.	Wulder et al. (2000) Culvenor (2002) Pouliot (2002), Wang et al. (2004)	Appropriate for conifers, with a conical shape and high reflectance point at the top of the tree. Simple method to use. Results are affected by the spatial distribution of trees, variation of tree crowns size, search window size (increased omission errors in larger windows and commission errors in smaller windows).
Template Matching (TM)	Pattern recognition	Quantitative descriptors used to describe patterns. Calculate the correlation between the image and the model. Model may be a sample extracted from the image or not.	Pollock (1996) Larsen (1997) Larsen and Rudemo (1998) Quackenbush et al. (2000) Erikson (2004) Hung et al. (2012)	Enables analysis of the tree crown from its spectral, textural and structural characteristics. Allows neighborhood analysis of the tree crown by considering its shadow. User-friendly method. Needs a template library, making it unpractical in complex forests. Recognition errors increase with irregularity of the tree crowns. Easier to detect larger trees than smaller ones. Performance reduced in very dense environments.
(VF)	Delineation of tree crown	Derives from an analogy with a topographical surface, programmed to identify the shaded portion between the tree crowns (valleys).	Gougeon (1995, 1999) Leckie and Gougeon (1998) Gougeon and Leckie (2003) Erickson (2004) Gougeon and Leckie (2006)	Appropriate for trees with conical shape that create shadow areas between individuals. Most successful to delineate populations of the same age without intertwined tree crowns. Best performance for images in mid-low solar elevation angle. Performance reduced when trees are asymmetrical, of from different species, with different tree crown sizes or when shadows of trees protrude over each other. Tendency to group smaller trees together and split larger trees into multiple segments.

Algorithm	Usage	Principle	Researches	Characteristics / Limitations
Watershed (WS)	Delineation of tree crown	Performed from the image gradient. Image is seen as an inverted topographic surface flooded to determine watershed divides. Commonly uses markers to limit the number of segments.	Wang *et al.* (2004) Lamar *et al.* (2005)	Performs best when applied after selection of markers to control segmentation process. More suitable for conifers, which allow preselection of treetops by using another approach (usually LM). Over-segmentation occurs when applied directly to the image or without the use of markers. Can separate tree crowns in different segments when the branches are too spread, or may include several trees in the same segment when there is no spatial separation between them.
Region Growing (RG)	Delineation of tree crown	Groups pixels or sub-regions based on predefined criteria for the growth of region in order to separate and recognize objects in the image.	Culvenor (2002) Pouliot et al. (2002) Erikson (2004) Bunting and Lucas (2006) Pu and Landry (2012)	More complex shapes of trees are better delineated. Method more complex as it requires different rules for different environments. Tends to create more than one segment when the tree has branches with dark portions, and tends to group different trees if they are very similar.
Marked Point Processes (MPP)	Pattern recognition	Stochastic process in which unordered points in a space are provided with marks. Marks are modeled from geometric and radiometric characteristics of objects.	Perrin (2006) Zhou (2012) Larsen *et al.* (2012)	Performs best with plantations of trees of same species and age and in images of isolated trees. It is less effective to detect trees in more complex environments.

Table 2. Summary of Local Maxima, Template Matching, Valley Following, Watershed, Region Growing and Marked Point Processes algorithms used to individual tree crown detection. The principles, main researches and main characteristics and limitations are presented.

Reference [11] compared six different algorithms (valley-following, region growing, template matching, scale-space theory, marked point processes, and Markov random fields) in six different aerial images, ranging from a homogeneous plantation and an area with isolated tree crowns to an extremely dense deciduous forest type. The authors found that none of the algorithms can by itself reach a high rate of success in all of the tested images and concluded that there is no single optimum algorithm for all types of images and forests. They also emphasized that for complex types of forests, monoscopic images are insufficient for a consistent detection of tree crowns, even for human interpreters.

3. Part II – A hybrid approach integrating marked point process and template matching

As shown in our brief review, many methods have been developed for trees in temperate forest environments. In an exploratory research [23], three algorithms in urban tropical environments were tested: region growing, watershed, and template matching. Better results were generally obtained by combining region-growing segmentation and geographic object-based image analysis (GEOBIA) for classification. Although highly effective, the approach requires much parameter setting and experience and is not especially dedicated to the problem of tree crown detection.

Studies that use marked point processes have triggered our attention and made us consider that they could benefit from using marks modeled from 3D objects in a different approach than from that developed by reference [5]. We propose to use a geometrical-optical tree model in a manner resembling that of template matching that uses some form of correlation between image and model to identify candidate pixels. An MPP taking advantage of a geometrical optical 3D model and measurements of similarity to seek tree crowns could represent a significant improvement to using simpler marks. Considering such a hypothesis, we developed an algorithm for tree crown detection that combines elements from MPP, TM, and tree crown geometrical-optical modeling for the automatic detection and (simplified) delineation of trees in VHR satellite imagery. We have named our algorithm MPP–TM.

In our approach, the TM did not scan the whole image like it was initially conceived but rather uses an MPP approach to select random locations within the image. Additionally, the 3D marks receive a random diameter between a predetermined range depending on the type of environment. The geometrical-optical model includes both the sunlit and shaded areas of the crown and a portion of the projected shadow to allow a better match between model and image. Some statistical and spectral parameters were also included in the model-matching phase.

MPP-based algorithm for pattern recognition usually alternates between phases of birth and death during which the objects are created (placed) and destroyed when they do not comply with the matching rules. This is also a characteristic of MPP–TM, but we have somewhat deviated from the original concept where the destruction phase also incorporated a random process.

The following subsections are devoted to describe the construction process of the 3D geometrical-optical model and the functioning of the algorithm.

3.1. Description of MPP–TM approach

3.1.1. A geometrical-optical 3D tree crown model

The parameters that determine the radiance pattern of a tree crown are direct and indirect radiation, shape of tree, branch pattern, leaf reflectance, multiple reflectances within the canopy, etc. [36]. In creating a valid 3D geometrical-optical model, we have chosen a simplified version in which the crown is represented by a dome of varying skewness, a Lambertian reflectance model with ambient light, and a projected shadow on the ground (or on another tree). Equations 2 and 3 give the formulation of our model in which each pixel is treated as a singular surface.

$$\cos(\theta_i) = \left\{\cos(\theta_s)\cos(\theta_n) + \sin(\theta_s)\sin(\theta_n)\cos(\varphi_s - \varphi_n)\right\} \tag{2}$$

where θ_i is the local solar incidence angle, θ_s is the solar zenith angle, θ_n is the slope of the object surface, φ_s is the solar azimuth and φ_n is the aspect of the object surface.

$$L = L_M \left(\frac{\cos(\theta_s)}{\cos(\theta_i)} \right) + \text{amb} \tag{3}$$

where L_M is the maximum reflectance of the model, "amb" represents the diffuse ambient lighting. The geometrical-optical model is adjusted according to the specific illumination parameters of the image, and the size of the trees present on the scene. Figure 7 shows two examples of tree models with similar reflectance but different solar elevations.

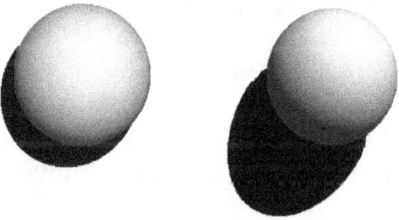

Figure 7. Illustration of the geometrical-optical model of tree crown as seen in the same sun azimuth (32°) but in two different solar elevation angles: (left) 20° and (right) 45°

A parameter of projected shadow clipping has also been added to account for the fact that it was not beneficial to use the whole shadow in situations where it was projected onto another tree and not on the ground. The height of the tree also affects the size of the shadow so that it did not appear wise to set the height to a fixed value. To illustrate this, Figure 8 shows a comparison between the tree model and an actual tree from the image both with whole and clipped shadows.

3.1.2. Algorithm description

According to reference [32], using MPP to extract objects consists in searching for the "best" possible object configuration in a scene, the one that will respect a certain number of properties

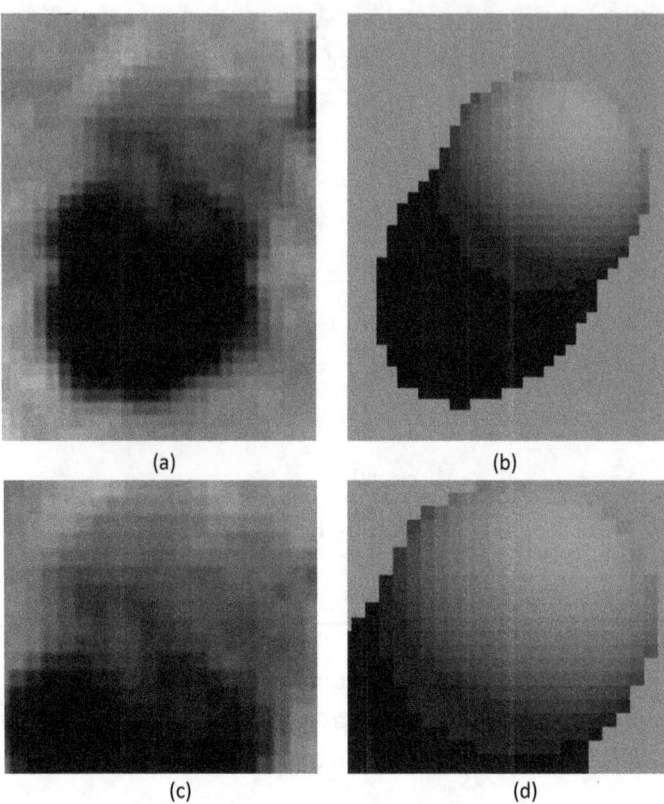

Figure 8. Comparison between an isolated tree from the (a) WorldView-2 image and (b) the geometrical-optical 3D model. A clipping factor of about 80% was applied to the same images in (c) and (d) to enable the use of only a portion of the shadow in cases where that shadow is not projected on the ground but on another object.

both of the objects being sought and the radiometric properties of the image. In our algorithm, the "best" configuration be it geometric or radiometric is given by the model.

The process consists in alternating phases of birth and death. The MPP starts with a birth phase during which tree crowns represented by circles of varying size (a randomized interval) are inserted on a matrix of equal size to the image being processed. Tree crowns are only inserted where no other crowns are present. Once all the circles have been inserted (determined by a density parameter N_c), a similarity (Sm) value between the image and a version of the model fitted on each circle is computed and stored in a list along with the parameters of the model. A routine then sorts the list by decreasing values of Sm. During the death phase, the circles that do not comply with the acceptance restrictions (Sm and minimum and maximum standard deviation threshold) are successively deleted from the matrix and the list. At the end of each death phase, the overall parameters of the pixel distribution of the remaining crowns are updated. The crowns that have been found are definitively kept but are re-thrown in the bundle of crowns of the next iterations. If after an iteration 10 crowns are kept out of 100, then the next iteration will randomly place 90 more crowns, and the new set of 100 crowns are evaluated and sorted for the next iteration. All tree crowns are considered "found" when one of the three possibilities is encountered: 1) the number of trees found is equal to the number given by the density parameter, 2) one of the interruption criteria has been attained, or 3) the maximum number of iteration has been reached.

The *Sm* value is computed as the subtraction of two parameters: cross-correlation and the normalized absolute difference as defined by the following relation (Equation 4):

$$Sm = \gamma - \alpha\,ND \tag{4}$$

where γ is the cross-correlation between image and model, ND is the normalized sum of absolute differences between them, and α is a constant weight factor (normally approx. 0.5). The cross-correlation and absolute difference are calculated as follows (Equations 5 and 6).

$$\gamma(x,y) = \frac{\sum_i \sum_j \left[w_{(i,j)} - \bar{w} \right] \sum_i \sum_j \left[f_{(i,j)} - \bar{f} \right]}{\left\{ \sum_i \sum_j \left[w_{(i,j)} - \bar{w} \right]^2 \sum_i \sum_j \left[f_{(i,j)} - \bar{f} \right]^2 \right\}^{1/2}} \tag{5}$$

In Equation 5, the cross-correlation is calculated between the model matrix ($w_{(i,\,j)}$) and the portion of the image that corresponds to the circle of the same radius ($f_{(i,\,j)}$). In other words, two matrices of same dimensions are always compared. $\bar{w}$ and $\bar{f}$ are their respective means. The values of gamma range between −1 and 1. The same logic is used in Equation 6 which computes a normalized difference value between the same two matrices.

$$ND = \frac{\sum w - \sum f}{\sum w + \sum f} \tag{6}$$

In the death phase, tree crowns are kept if their similarity is larger or equal to a pre-set threshold. Because we found that such a threshold represented a weak element in our algorithm, we implemented a strategy by which it needs not be predetermined with a fixed value but rather adjusts itself as the number of iterations grows. The threshold is set very high at the beginning but then starts to decay when a certain number of iterations do not find any "new" tree crown (typically 100 iterations). Additionally, if more than a certain amount of iterations (say 1000) still does not add any new tree crown, then the process is stopped. Ultimately, it will be stopped if the maximum number of iterations is reached. A flowchart of our algorithm is presented in Figure 9 and schematically described in Table 3.

Figure 10a shows the state of the crown matrix after a single birth phase with 163 circles of random radius (between 3 and 15 m) and randomly located within the image matrix. After the death phase, using a similarity threshold of 0.98, only one tree crown was kept (Figure 10b).

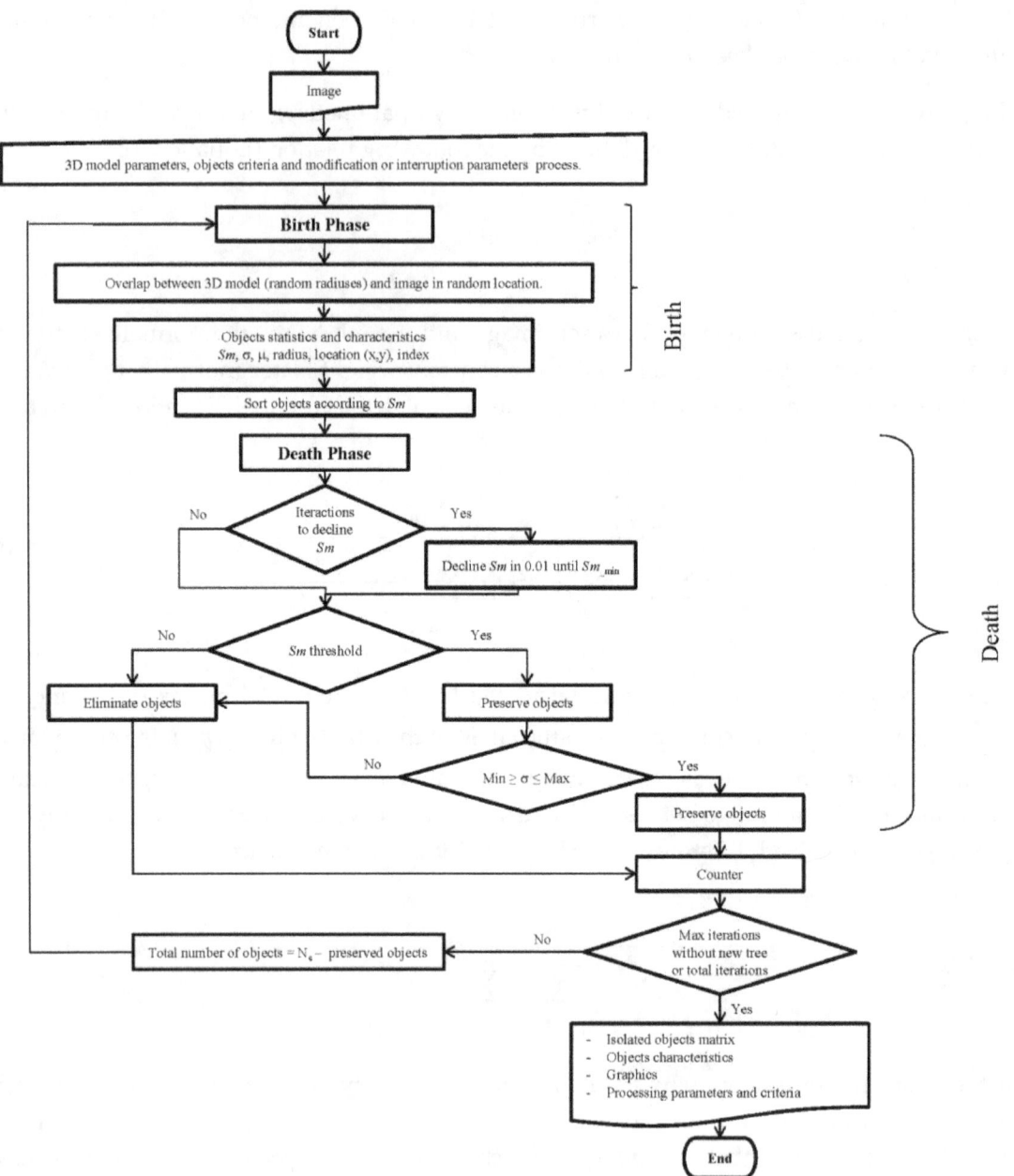

Figure 9. Flowchart of the MPP–TM algorithm.

3.1.3. A modified approach for orchards

Because trees in orchards are often individually distinguishable and have similar shape and size, they are perfect candidates for TM with a 3D geometrical-optical model. By using a GOM, the effects of varying illumination (sun elevation and azimuth) become an advantage rather than an obstacle especially when the background is homogeneous. In terms of data, VHR image data such as a large proportion of Google Earth images have sufficient resolution for identi-

1. Task: Tree crown detection in Very High Resolution images.

2. Set parameters:

a. 3D model: maximum reflectance, ambient light, sun elevation, sun azimuth, tree shape, clip factor.

b. Descriptors of the objects: minimum and maximum radius, minimum and maximum standard deviation (δ) threshold, maximum and minimum similarity(sm) threshold, trees density.

c. Change the process: maximum iterations for decrease similarity

d. Interruption of the process: total iterations, maximum iterations without find new trees.

3. Approach to tree crown detection:

a. While the number of searched trees is not achieved or some of the interruption process (total iterations or minimum threshold for similarity).

4. Starts the birth phase:

a. Randomly pick a radius within model catalogue

b. Randomly pick i and j coordinates within the image space

c. Check if crown is already present

d. If not:

i. Fill area with circle of radius r

ii. Extract corresponding area in the image matrix

iii. Compare image and model matrices

iv. Calculate and store values: i,j, average, standard deviation and Sm

5. Starts the death phase:

a. Input parameters: birth image matrix; crown statistics (Sm sorted); Sm threshold; tree models catalogue with radius between maximum and minimum radius

b. While smcrown < smthreshold:

i. Zero crown pixels in birth image matrix

c. While δ crown < min δ threshold and δ crown > max δ threshold:

i. Zero crown pixels in birth image matrix

6. Update object and global statistics

7. Update number of crowns eliminated for next birth phase

8. When the process finish: reports, graphs and image with individual tree crowns

Table 3. Description of MPP-TM algorithm.

fying orchard trees. In this case, however, illumination parameters are not readily available and must be determined.

The objective of this modified approach is to introduce an adaptation of the algorithm described earlier to detect and count trees in orchards of different types. Because it was aimed at a more regional or even global application, Google Earth images were used in an attempt to simulate a generic operational framework. The modified approach uses a similarity measurement between the GOM and the image to calculate the probability of being the center of tree and then places trees in nonoverlapping positions (unless some overlapping is allowed). The algorithm also incorporates a module to determine the illumination parameters from a sample.

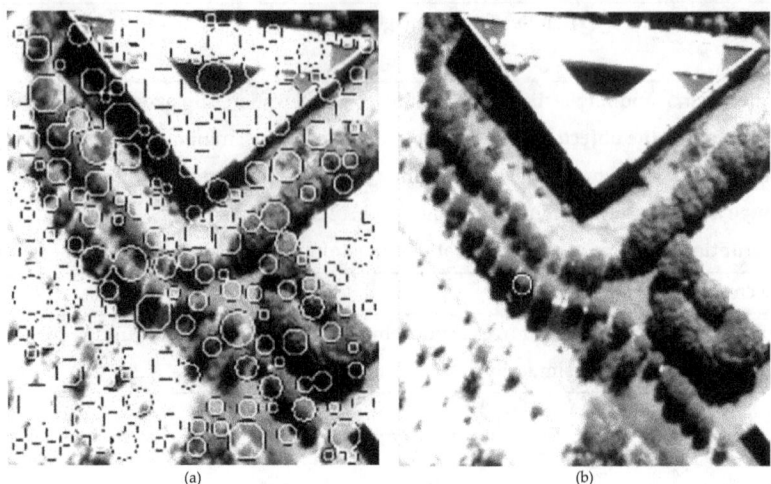

(a)   (b)

Figure 10. Illustration of the (a) birth and (b) death phases of the MPP–TM algorithm. In this example, of the 163 randomly positioned crowns, only one had a similarity value larger than the threshold of 0.98.

The algorithm is based on three principles. First, it assumes that the trees have a dome-like shape approximated with a GOM and the right illumination parameters. Second, there is little or no overlapping between trees, and third, the pixel with the highest similarity represents the most likely central position of the tree.

The GOM is a simple dome model for which the height is estimated at 1.5 times the diameter of the crown, and to simplify the problem we have assumed a unique diameter for all trees in the orchard (this can easily be modified to incorporate a range of diameters). The algorithm responsible for the detection of trees are best explained through a list of steps.

Step 1. Get user parameters: percent overlapping allowed minimum similarity value, tree diameter, and coordinate of sample tree. These parameters cannot be estimated automatically and are entered by the user. The illumination parameters can optionally be entered by the user, else they will be estimated by the program using the tree sample.

Step 2. If sun elevation and azimuth are not provided by the user, the parameters are automatically estimated by the program using the coordinates of a single-tree sample. The program then computes the similarity between the sample and all possibilities of illuminations parameters in steps of 10 degrees.

Step 3. Calculate the similarity value for each pixel.

Step 4. Sort pixels by decreasing similarity and store coordinates. If the value is lower than the minimum allowed, the pixel is not stored.

Step 5. Place a temporary tree "stamp" (flat template) at the next pixel location with highest similarity value.

Step 6. Verify if space is already occupied by a tree. If some overlapping is allowed, make sure that the number of nonzero pixel is smaller than the percentage of overlapping allowed. An output image is created to receive a permanent "stamp" of the GOM shape with the value of similarity associated.

Step 7. Validate the results. Validation is performed by estimating the overall number of trees using the density of a representative sample and comparing with the number of trees found.

3.2. Testing the MPP–TM approach

3.2.1. Urban trees

Urban trees play an important role in the welfare and quality of life in cities. They contribute to improving air and water quality, mitigate the carbon dioxide and other pollutants, moderate the microclimate and air temperature, help control soil erosion, reduce the flow of rainwater, and provide biodiversity [37–39]. A good knowledge of the species planted in cities and their health contributes to the inventory and management of these trees. To fulfill their role in the urban environment, trees need to be looked after through maintenance practices such as pruning and monitoring them for pests and diseases.

A WorldView-2 (WV-2) image of the campus of the Universidade Federal de Minas Gerais (UFMG) (and surroundings) in Belo Horizonte, Brazil, was used as our test data (Figure 11). The scene was already orthorectified and radiometrically corrected. Although WV-2 offers nine different spectral bands, only the panchromatic band ($\lambda \approx 450$–800 nm) with a ground resolution of 50 cm was used since all other bands have a ground resolution of 2 m.

Three WV-2 sub-images were selected to test the performance of MPP–TM algorithm (Figure 10). These images were chosen from different contexts with both isolated and grouped trees and with other objects present in the scene. A wide variety of crown radii is also present in these images. The first two images (Figure 12a and b) are from the university campus of UFMG, and the last is from an urban park (Figure 12c).

To assess the quality of the results produced by MPP–TM, validation was done by comparing our results with a visual interpretation of the trees in the image. For these, only tree counting was used as validation. For the crown counting validation, we considered the following situations: 1) true positives (TP) for found trees, 2) false positives (FP) when a detected object is not a tree, and 3) false negatives (FN) for trees not encountered. The success score was computed as follows (Equation 7):

$$A = \left(\frac{TD - FP}{N + FN} \right) \times 100 \tag{7}$$

where TD represents the total detected trees and N is the total number of trees.

These results are shown below with their respective overall similarity and standard deviation graphs (Figure 13). The validation results are presented in Table 4.

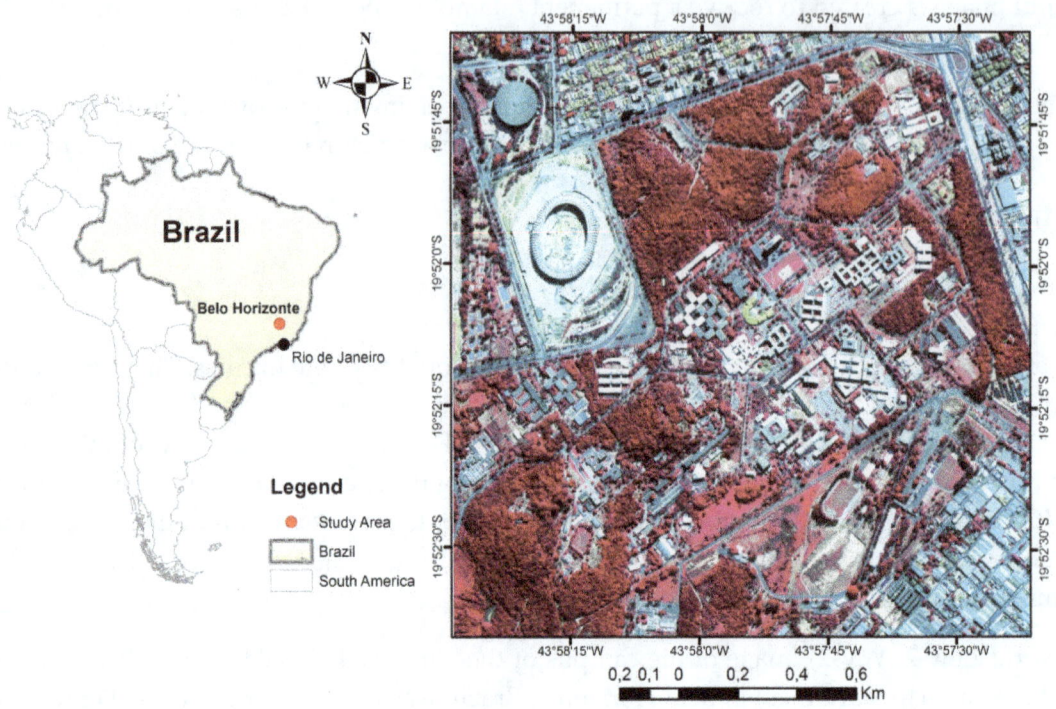

Figure 11. Location of study area. The image on the right is a WorldView-2 false color composite

Image	Number of Trees	Trees Detected	False Positive	False Negative	Overall Accuracy (%)
WV image 1	47	43	3	8	72.73
WV image 2	50	59	8	5	92.73
WV image 3	175	161	5	20	80.00

Table 4. Validation of the MPP-TM results with the three WV-2 images.

In the two images of the campus, the program was able to find 73% and 93% of the trees, respectively, with very few errors in isolated trees (Figure 13a and b). The presence of other objects (buildings, streets, and sidewalks) did not hinder the identification of trees and few false positives (3 and 8, respectively) were found. In both images, MPP–TM was able to find most grouped trees, but the crown diameter was often slightly off. It should be noted that some cases are even difficult to correctly identify and delineate visually. Mostly, the errors came from dividing a single crown into two, or including two different crowns as a single object.

The WV-2 image 3 is from a protected urban park area with predominantly isolated trees and relative homogeneous crown size of about 6 m (Figure 13 c). A total of 161 objects were detected with only 5 false positives and 20 false negatives for an overall success of 80%. Although most deciduous trees were selected, the crown size was often incorrect but given the highly irregular

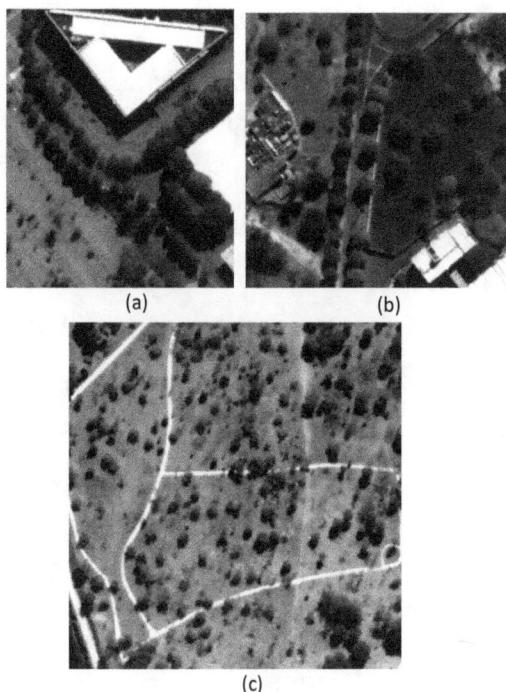

Figure 12. Sub-images selected from the WV-2 image.

shape of many of these trees, this was somewhat expected, and similar problems have been reported by reference [4].

The behavior of the overall similarity during the iterations tend to increase as the image is progressively occupied by found trees and this is why the overall similarity increases. The standard deviation, however, is very different for each image and is mostly related to the amount of contrast in the original image. Images with highly contrasting objects (e.g., building tops) will tend to show a progressively decreasing standard deviation. Images of low contrast will tend to see it increasing as the trees are progressively added because of the double illumination nature of the trees.

3.2.2. Orchards

Orchards are collections of individual trees often arranged regularly for which the MPP–TM algorithm could easily be adapted. Tree counting in orchards can be very useful for inventory and management purposes. For instance, the European Union (EU) Common Agricultural Policy (CAP) regulations (EC 73/2009) provide support for permanent crops such as hazelnuts, almonds, walnuts, and fruits in general [40–42]. Eligible orchards need to have a certain size and tree density depending on the type of crop. It has been estimated that orchard fruit production represents approximately 3– 4% of the total arable land [43], so the task of esti-mating fruit production needs tools for counting trees in a timely fashion. Furthermore, the task can take advantage of the near-global high-resolution image cover provided by Google (Google Earth and Google Map) and other Internet-based image services.

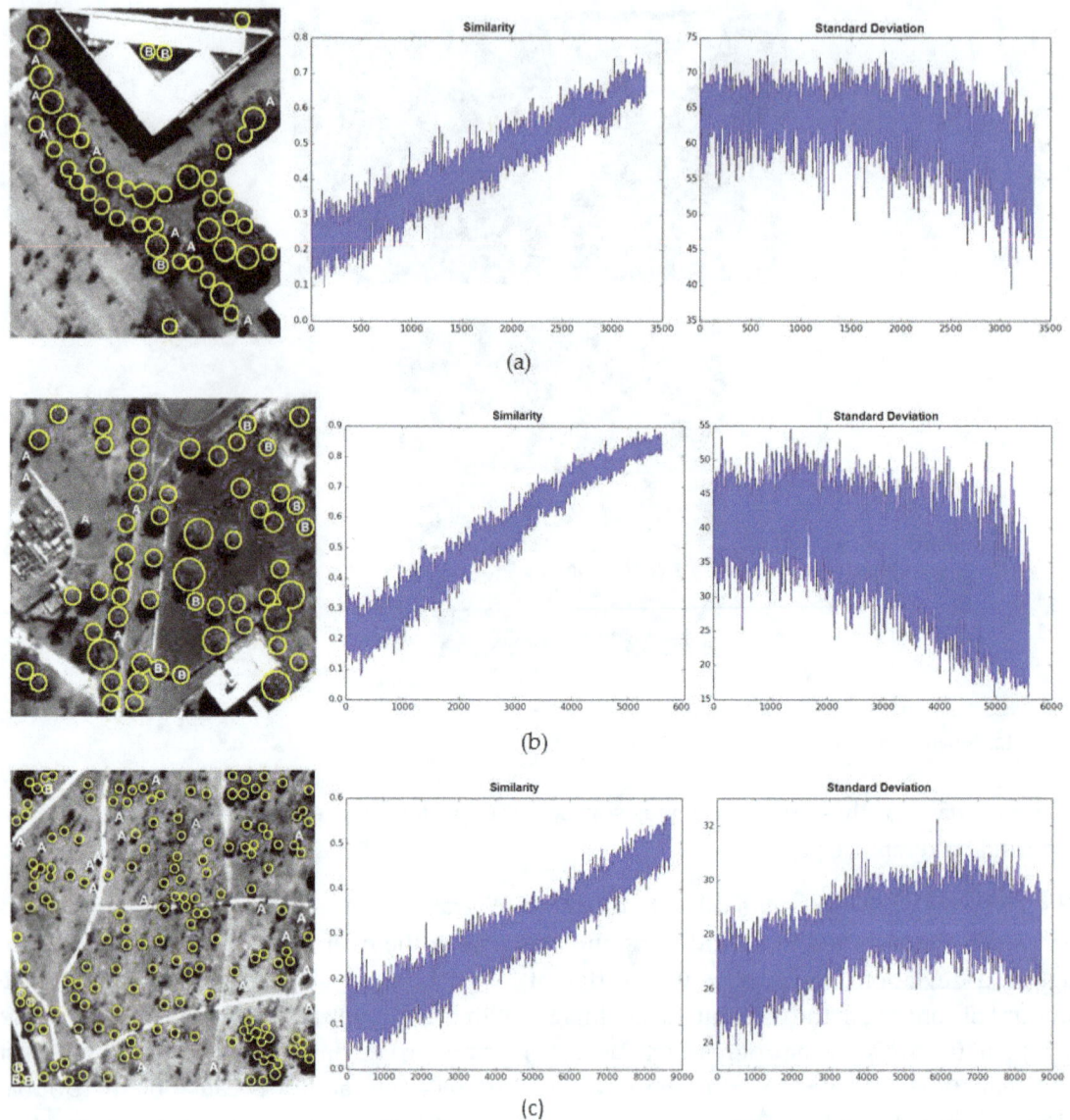

Figure 13. MPP–TM results obtained with the three WV-2 image 1–3 (left) and their graphs of global similarity (center) and standard deviation (right). The yellow circles correspond to correctly identified trees (true positive or TP), the objects marked with a yellow "A" are false negatives (FN) and the objects marked with a yellow "B" are false positives (FP).

Orchards are plantation of trees of the same species and often of the same age. Consequently, trees of orchards usually have similar size and shape and are regularly spaced. Image processing can easily be adapted to such a task providing VHR images are available. To illustrate the adapted MPP–TM algorithm (which no longer is a real MPP), we have tested over three different types of orchards: a mango plantation in Brazil near Juazeiro, a walnut plantation in France near Grenoble, and an olive plantation in Italy near Bracciano. The three images were directly extracted from Google Earth and had a relatively bad quality as they appeared to have been enhanced for sharpness. To validate the results, we have asked three geography students to manually interpret and mark the trees belonging to orchards for the three test images, and we have evaluated the results in the following way:

- the total number of trees (NT) was determined by the interpreters;

- matched trees were computed as true positive and are defined by the number of trees found by the algorithm minus the false positives;

- unmatched trees (present on the image but absent from the results) were computed as false negative (FN);

- trees marked by the algorithm but not by the interpreters were marked as false positive;

- the final accuracy was computed as TP / (NT + FN).

To be fair, the interpreters were told not to mark the trees that seem too small or too big for the orchards. In addition, valid trees that were found by the algorithm but did not pertain to an orchard were not computed as false positive. As a further improvement, restricting the search within the boundaries of the orchards would increase the accuracy and enable the similarity parameter to be relaxed. The addition of other spectral bands should also improve the results.

Test Image	Number of trees	True positives	False positives	False negatives	Overall accuracy
Grenoble	2435	(2358-69)= 2289	69	103	90.19%
Bracciano	837	(1071-264)=807	264	29	93.19%
Juazeiro	2534	(2555-114)= 2441	114	93	92.92%

Table 5. Results of the tree counting algorithm for the three regions (France, Italy and Brazil).

Table 5 shows an overview of the results for the three test images, and Figure 14 shows the graphical results. The top row shows the original images, the center row shows the results of the tree identification (as well as false positives and negatives), and the bottom row displays a detailed section of the image on which the results were overlaid. The Grenoble test image (Figure 14 left column) was characterized by densely arranged walnut trees, which have a large round crown so that the model was well correlated with trees on the image, but the fact that the trees are close to one another produced a relatively large number of "miss" (103). This forced to relax the similarity threshold and caused a few false positives (69). In the case of the Bracciano image (Figure 14 center column), the olive trees are more ill- shaped than the walnut trees, and the relaxation of the similarity threshold caused a large number of false positives, especially in the nearby forested areas. Conversely, very few trees were missed. Finally, the last test image from Juazeiro (Figure 14 right column) is populated by mango trees that, like the walnut trees, have large round crowns. Still, the algorithm produced a fair amount of both false positives and false negatives mainly because of the variation of tree crown size and the particular situation of the dirt road at the top of the image that created a pattern of light and shade similar to the trees (approximately one-third of the false positives came from that road). The three very different images still produced similar accuracy results between 90 and 93%.

Figure 14. Illustration of the results of the tree counting for the three test images: Grenoble (left column), Briacciano (center column), and Juazeiro (right column). The empty circles represent the trees that were found, "x" represents the false positives and the black circles represent the false negatives.

4. Final considerations

The detection of individual tree crown in images of very high resolution is a growing and challenging field of research within the remote sensing community. In addition to the structural complexity of the forest, many other factors such as the characteristics of the scene (topography, illumination, and other environmental variables) and forest type (season and biodiversity) make the task difficult. To reference [16], the ability to achieve individual tree crown delineation of all trees in a forest was recognized as an unrealistic expectation.

In an effort to provide the reader with an overview of the current state of the research in tree crown detection, Part I presented a brief review of some of the most common computerized techniques for detecting and delineating trees in optical VHR images. Part II describes the concepts and implementation of a novel approach based on two mathematical/pattern recognition concepts integrated to improve performance. MPP–TM was developed based on concepts from marked point processes and template matching for the former to take advantage of a mark built from a geometrical-optical model.

MPP–TM was highly effective in finding trees in urban environment with images from the WorldView-2 satellite (ground resolution of 50 cm). A total of 263 trees out of 272 were found (96%), and taking false positives into account, a success rate over 90% was still achieved. The algorithm was also adapted for a tree counting application such as is often needed in large orchards. To count trees in orchards, the approach works very well when the trees are easily distinguishable. Results from three datasets of different crops show an average success better than 90%. Out of 5806 trees, 5537 were found excluding all false positives.

The growing availability of VHR images from commercial satellites or even from web mapping services opens a wide field of applications especially that VHR multispectral images are becoming increasingly common. Multi-temporal studies will further strengthen these applications for monitoring purposes.

Finally, we should mention that Lidar (light detection and ranging) data are also becoming widely available, and its integration with VHR images promises to further improve the results of tree detection algorithm. By adding a third dimension to the images, Lidar reduces the probability of errors by strengthening the evidence around the digital representation of trees.

Acknowledgements

We are grateful to François Gougeon, Donald Leckie, Guillaume Perrin and Mats Erikson for having kindly provided the rights of reproduction of their figures.

Author details

Marilia Ferreira Gomes[1,2*] and Philippe Maillard[1]

*Address all correspondence to: mariliafgomes@yahoo.com

1 Geography Department, Geosciences Institute, Federal University of Minas Gerais, Belo Horizonte, Brazil

2 Cartography Department, National Institute of Land Reform, Belo Horizonte, Brazil

References

[1] Pollock RJ. Model-based approach to automatically locating tree crowns in high spatial resolution images. In: Jacky Desachy, editor. Proceedings SPIE 2315: Image and Signal Processing for Remote Sensing; 30 December 1994; Rome. International Society for Optics and Photonics; 1994. p. 526–537. DOI: 10.1117/12.196753

[2] Avery TE, Berlin GL. Fundamentals of remote sensing and airphoto interpretation. 5th ed. New Jersey: Prentice Hall; 1992. 472 p.

[3] Shao G, Reynolds KM, editors. Computer applications in sustainable forest management: Including perspectives on collaboration and integration. 1st ed. Dordrecht: Springer Netherlands; 2006. 277 p. DOI: 10.1007/978-1-4020-4387-1

[4] Ke Y, Quackenbush LJ. A review of methods for automatic individual tree-crown detection and delineation from passive remote sensing. International Journal of Remote Sensing. 2011;32(17):4725–4747. DOI: 10.1080/01431161.2010.494184

[5] Perrin G. Etude du couvert forestier par processus ponctuels marqués [thesis]. Paris: Ecole Centrale Paris; 2006. 170 p. Available from: https://tel.archives-ouvertes.fr/tel-00109074/ [Accessed: 2015-06-08]

[6] Erikson M. Segmentation and classification of individual tree crowns in high spatial resolution aerial images [thesis]. Uppsala: Swedish University of Agricultural; 2004. 45 p. Available from: http://pub.epsilon.slu.se/676/[Accessed: 2015-04-12]

[7] Gougeon FA, Leckie DG. Forest information extraction from high spatial resolution images using an individual tree crown approach. 1st ed. Victoria: Canadian Forest Service; 2003. 27 p.

[8] Franklin SE. Remote sensing for sustainable forest management. 1st ed. New York: CRC Press; 2001. 424 p.

[9] Blaschke T. Object based image analysis for remote sensing. ISPRS Journal of Photogrammetry and Remote Sensing. 2010;65(1): 2–16. DOI: 10.1016/j.isprsjprs.2009.06.004

[10] Pu R, Landry S. A comparative analysis of high spatial resolution IKONOS and WorldView-2 imagery for mapping urban tree species. Remote Sensing of Environment. 2012;124:516–533. DOI: 10.1016/j.rse.2012.06.011

[11] Larsen M, Eriksson M, Descombes X, Perrin G, Brandtberg T, Gougeon F. Comparison of six individual tree crown detection algorithms evaluated under varying forest conditions. International Journal of Remote Sensing. 2011;32(20):5827–5852. DOI: 10.1080/01431161.2010.507790

[12] Hay GJ, Castilla G. Geographic Object-Based Image Analysis (GEOBIA): A new name for a new discipline?. In: Blaschke T, Lang S, Hay G, editors. Object-Based Im-

age Analysis - Spatial Concepts for Knowledge-Driven Remote Sensing Applications. 1st ed. Berlin: Springer; 2008. p. 1.4. DOI: 10.1007/978-3-540-77058-9

[13] Wulder M, Niemann KO, Goodenough DG. Local maximum filtering for the extraction of tree locations and basal area from high spatial resolution imagery. Remote Sensing of Environment. 2000;73(1):103–114. DOI: 10.1016/S0034-4257(00)00101-2

[14] Pouliot DA, King DJ, Bell FW, Pitt DG. Automated tree crown detection and delineation in high-resolution digital camera imagery of coniferous forest regeneration. Remote Sensing of Environment. 2002;82(2):322–334. DOI: 10.1016/S0034-4257(02)00050-0

[15] Brandtberg T, Warner T. High-spatial-resolution remote sensing. In: Shao G, Reynolds KM, editors. Computer Applications in Sustainable Forest Management: Including Perspectives on Collaboration and Integration. 1st ed. Dordrecht: Springer Netherlands; 2006. p. 19–41. DOI: 10.1007/978-1-4020-4387-1

[16] Culvenor DS. TIDA: an algorithm for the delineation of tree crowns in high spatial resolution remotely sensed imagery. Computers & Geosciences. 2002;28(1):33–44.

[17] Gonzales RC, Woods RE. Digital Image Processing. 3rd ed. New Jersey: Pearson Prentice Hall; 2008. 954 p.

[18] Caves RG, Harley PJ, Quegan S. Matching map features to synthetic aperture radar (SAR) images using Template Matching. IEEE Transactions on Geosciences and Remote Sensing. 1992;30(4):680–685. DOI: 10.1109/36.158861

[19] Pollock RJ. The automatic recognition of individual trees in aerial images of forests based on a synthetic tree crown image model [thesis]. Toronto: The University of British Columbia; 1996. 172 p. Available from: https://circle.ubc.ca/handle/2429/6135

[20] Larsen M. Crown modelling to find tree top positions in aerial photographs. In: In Proceedings of the Third International Airborne Remote Sensing Conference and Exhibition, volume II; 7-10 July 1997; Copenhagen. Ann Arbor: ERIM International; 1997. pp. 428–435.

[21] Larsen M, Rudemo M. Optimizing templates for finding trees in aerial photographs. Pattern Recognition Letters. 1998;19(12):1153–1162. DOI: 10.1016/S0167-8655(98)00092-0

[22] Quackenbush LJ, Hopkins PF, Kinn GJ. Developing forestry products from high resolution digital aerial imagery. Photogrammetric Engineering and Remote Sensing. 2000;66(11):1337–1346.

[23] Gomes MF, Maillard P. Identification of urban tree crown in a tropical environment using WorldView-2 data: problems and perspectives. In: Michel U, Civco DL, Schulz K, Ehlers M, Nikolakopoulos KG, editors. SPIE 8893, Earth Resources and Environmental Remote Sensing/GIS Applications IV; 23–25 September; Dresden. Washington: SPIE; 2013. p. 88930C-88930C-13. DOI: 10.1117/12.2029073

[24] Hung C, Bryson M, Sukkarieh S. Multi-class predictive template for tree crown detection. ISPRS Journal of Photogrammetry and Remote Sensing. 2012;68:170–183. DOI: 10.1016/j.isprsjprs.2012.01.009

[25] Gougeon F. A crown-following approach to the automatic delineation of individual tree crowns in high spatial resolution aerial images. Canadian Journal of Remote Sensing. 1995;21(3):274–284.

[26] Leckie DG, Gougeon FA. An assessment of both visual and automated tree counting and species identification with high spatial resolution multispectral imagery. In: Hill DA, Leckie DG, editors. International Forum on Automated Interpretation of High Resolution Digital Imagery for Foresty; 10–12 February 1998; Victoria. Victoria: Canadian Forest Service; 1998. p. 141–154.

[27] Gougeon FA, Leckie DG. The individual tree crown approach applied to Ikonos images of a coniferous plantation area. Photogrammetric Engineering & Remote Sensing. 2006;72(11):1287–1297. DOI: 10.14358/PERS.72.11.1287

[28] Szeliski R. Computer vision: algorithms and applications. 1st ed. London: Springer-Verlag; 2010. 812 p. DOI: 10.1007/978-1-84882-935-0

[29] Wang L, Gong P, Biging GS. Individual tree-crown delineation and treetop detection in high-spatial-resolution aerial imagery. Photogrammetric Engineering & Remote Sensing. 2004;70(3):351–357. DOI: 10.14358/PERS.70.3.351

[30] Lamar WR, McGraw JB, Warner TA. Multitemporal censusing of a population of eastern hemlock (Tsuga canadensis L.) from remotely sensed imagery using an automated segmentation and reconciliation procedure. Remote Sensing of Environment. 2005;94(1):133–143. DOI: 10.1016/j.rse.2004.09.003

[31] Bunting P, Lucas RM. The delineation of tree crowns in Australian mixed species forests using hyperspectral Compact Airborne Spectrographic Imager (CASI) data. Remote Sensing of Environment. 2006;101(2):230–248. DOI: 10.1016/j.rse.2005.12.015

[32] Zhou J. Application de l'identification d'objets sur images à l'étude de canopées de peuplements forestiers tropicaux: cas des plantations d'Eucalyptus et des mangroves [thesis]. Montpellier: Université Montpellier II - Sciences et Techniques du Languedoc; 2012. 191 p. Available from: https://tel.archives-ouvertes.fr/tel-00763706/

[33] Descombes X. Méthodes stochastiques en analyse d'image: des champs de Markov aux processus ponctuels marqués [thesis]. Nice: Université de Nice Sophia-Antipolis; 2004. 225 p. Available from: https://tel.archives-ouvertes.fr/tel-00506084/[Accessed: 2015-09-20]

[34] Ortner M. Processus ponctuels marqués pour l'extraction automatique de caricatures de bâtiments à partir de modèles numériques d'élévation [thesis]. Nice: Université de Nice Sophia-Antipolis; 2004. 250 p. Available from: https://tel.archives-ouvertes.fr/tel-00189803 [Accessed: 2015-09-01]

[35] Baddeley AJ, Van Lieshout MNM. Stochastic geometry models in high-level vision. Journal of Applied Statistics. 1993;20(5-6):231–256. DOI: 10.1080/02664769300000065

[36] St-Onge B. L'apport de la texture des images numériques de haute resolution a la cartographie forestière automatisée [thesis]. Montréal: Université de Montréal; 1994. 502 p.

[37] Östeberg J, Delshammar T,Wiström B, Nielsen AB. Grading of parameters for urban tree inventories by city oficials, arborists, and academics using the Delphi method. Environmental Management. 2013;51(3):694–708. DOI: 10.1007/s00267-012-9973-8

[38] Sander H, Polasky S, Haight RG. The value of urban tree cover: A hedonic property price model in Ramsey and Dakota Counties, Minnesota, USA. Ecological Economics. 2010;69(8):1646–1656. DOI: 10.1016/j.ecolecon.2010.03.011

[39] Almeida, A. L. B. S. S. S. L.. O valor das árvores: árvores e floresta urbana de Lisboa [thesis]. Lisboa: Instituto Superior de Agronomia; 2006. 342 p. Available from: http://hdl.handle.net/10400.5/469[Accessed: 2015-06-01]

[40] Santoro F, Tarantino E, Figorito B, Gualano S, D'Onghia AM. A tree counting algorithm for precision agriculture tasks. International Journal of Digital Earth. 2013;6(1): 94–102. DOI: 10.1080/17538947.2011.642902

[41] Aksoy S, Yalniz IZ, Tasdemir K. Automatic detection and segmentation of orchards using very high resolution imagery. Geoscience and Remote Sensing. 2012;50(8): 3117–3131. DOI: 10.1109/TGRS.2011.2180912

[42] Tasdemir K. Classification of hazelnut orchards by self-organizing maps," in Pattern Recognition in Remote Sensing (PRRS). In: Aksoy S, Younan NH, Forstner W, editors. IAPR Workshop on Pattern Recognition in Remote Sensing (PRRS); 22–22 August 2010; Istanbul. Institute of Electrical and Electronics Engineers IEEE; 2010. pp. 1–4. DOI: 10.1109/PRRS.2010.5742803

[43] Food and Agriculture Organziation of the United Nations. Statistical yearbook 2013: World food and agriculture. 1st ed. Rome: FAO; 2012. 289 p.

5

Bio-Optical Modeling in a Tropical Hypersaline Lagoon Environment

Igor Ogashawara, Marcelo P. Curtarelli, Carlos A. S. Araujo and José L. Stech

Additional information is available at the end of the chapter

Abstract

In this chapter, we attempted to present an overview of the use of remote sensing to monitor water quality parameters, mainly chlorophyll-*a* (chl-*a*) and turbidity. We summarized the main concepts of bio-optical modeling and presented a case study of the application of the Hyperspectral Imager for the Coastal Ocean (HICO) for the monitoring of water quality in a tropical hypersaline aquatic environment. Using HICO, we evaluated a set of different semi-empirical bio-optical algorithms for chl-*a* and turbidity estimation developed for inland and oceanic waters in the Araruama Lagoon, RJ, Brazil, which is an extreme environment due to its high salinity values. We also developed an empirical algorithm for both water quality parameters and compared the performances. Results showed that for chl-*a* estimation all models have a low performance with a normalized root mean square error (NRMSE) varying from 24.13 to 30.46. For turbidity, the bio-optical algorithms showed a better performance with the NRMSE between 15.49 and 28.04. Overall, these results highlight the importance of including extreme environments, such as the Araruama Lagoon, on the validation of bio-optical algorithms as well as the need for new orbital hyperspectral sensors which will improve the development of the field.

Keywords: Water quality, chlorophyll-*a*, turbidity, bio-optical modeling

1. Introduction

Earth Observations from space began in August, 1972, with the launch by National Aeronautics and Space Administration (NASA) of the Earth Resources Technology Satellite (ERTS-1) [1]. However, the use of remote sensing techniques to monitor inland water quality parameters such as chlorophyll-a (chl-*a*), total suspended solids (TSS) and turbidity only started to be extensively used in the past two decades with the development of bio-optical algorithms as well as the new hyperspectral and multispectral sensors. The use of optical remote sensing enables spatiotemporally comprehensive assessment of optical properties of the water column.

Water column optical properties are grouped into inherent optical properties (IOPs) and apparent optical properties (AOPs). IOPs are related to those properties that depend only upon the environment, thus, they are independent of the environment light field. The two most essential IOPs are the total absorption coefficient (a) and the total scattering coefficient (b) and the sum of both coefficients is the attenuation coefficient (k). AOPs, on the other hand, are those properties that depend on the environment and also on the directional structure of the environment light field. AOPs are also used as descriptors of a water body due to their regular features and stability. The most common AOPs are the irradiance reflectance (R), the remote-sensing reflectance (R_{rs}) and various diffuse attenuation functions [2]. A list of the most common IOPs and AOPs used in the literature is shown in Table 1.

Quantity	Units (SI)	Simbology
Inherent Optical Properties		
Absorption coefficient	m^{-1}	a
Volume scattering function	m^{-1} sr^{-1}	β
Scattering phase function	m^{-1}	$\tilde{\beta}$
Scattering coefficent	m^{-1}	b
Backscatter coefficient	m^{-1}	b_b
Beam attenuation coefficient	m^{-1}	c
Single-scattering albedo	-	ω_0
Apparent Optical Properties		
Irradiance reflectance (ratio)	-	R
Remote sensing reflectance	sr^{-1}	R_{rs}
Remote sensing reflectance (sub)	sr^{-1}	r_{rs}
Attenuation coefficients:		
of radiance L(z, θ, φ)	m^{-1}	$K(\theta, \varphi)$
of downwelling irradiance E$_d$(z)	m^{-1}	K_d
of upwelling irradiance E$_u$(z)	m^{-1}	K_u
of Photosynthetic Active Radiation (PAR)	m^{-1}	$KPAR$

Table 1. IOPs and AOPs commonly used in optical hydrology

Based on the interaction among AOPs and IOPs, absorption, scattering and attenuation properties of the water column are retrieved from proximal, aerial or orbital measurements of the solar spectrum mainly in the visible and near-infrared (NIR) spectral range. These optical properties allow the estimation of different water quality parameters such as: primary production, turbidity, eutrophication, particulate and dissolved carbon contents or the assessment of currents and algal blooms [3]. The relation among all these optical properties as well as the equipment to measure them were developed by oceanographers based on the modeling of downwelling solar and sky radiation spectra with the air–water interface and the subsurface aquatic absorption and scattering centers. Studies such as [4–8], among numerous others, established the main theory of the field before or around the launch of ERTS-1.

The first application of the theories of hydrologic optics was described by [9], which used a Monte Carlo simulation of the radiative transfer equation to relate the AOPs to the IOPs in oceanic waters containing optically active constituents, molecular water and chl-a. For inland waters, the first application of the hydrologic optics theories was developed for Lake Ontario, Canada, by a Monte Carlo simulation of the radiative transfer equation and non-linear multivariate optimization analyses [10]. These applications started a relatively new area for remote sensing applications known as bio-optical modeling, which focus on the use of the radiative transfer theory to derive optical properties or biological activity in the water column [2]. In [11], a classification of the bio-optical modeling products (algorithms) was proposed by describing five different types of algorithms: empirical, semi-empirical, semi-analytical, quasi-analytical and analytical. In this classification, the first two types (empirical and semi-empirical) and the last one (analytical) are usually used to estimate the biological activity from AOPs using more statistical methods, while the other two types (semi-analytical and quasi-analytical) are used to estimate IOPs from AOPs using the radiative transfer theory.

The development of bio-optical algorithms usually starts by collecting in situ limnological data as well as hyperspectral R_{rs} using a proximal sensor. The use of a hyperspectral sensor is appropriated to explore absorption peaks, which are very narrow to be identified by a multispectral one, to develop an algorithm. However, the ultimate goal for a bio-optical algorithm is to test its applicability on orbital sensors in order to become an important monitoring tool. Two of the most used satellite sensors to monitor water quality are the medium resolution imaging spectrometer (MERIS) and moderate resolution imaging spec-troradiometer (MODIS); both sensors provide the necessary spectral bands; however, their coarse spatial resolution makes them suitable only for very large aquatic systems. Despite the limitation on their spatial resolution, several research focus on the use of these two sensors for the monitoring of water quality parameters in inland water. Ref. [12] evaluated the perform-ance of different chl-a semi-empirical algorithms developed, especially for MERIS on a tropical reservoir in Brazil, and in [13] the authors developed a series of steps to improve the estimation of chl-a and cyanobacteria blooms in inland and near-coastal waters based on the MERIS imagery. For MODIS, empirical [14] and semi-empirical [15–17] algorithms have been developed for the monitoring of different water quality parameters.

To overpass the problem of the spatial resolution and to keep a good spectral resolution, hyperspectral airborne sensors have been used to monitor the quality of inland waters. One of the most common airborne hyperspectral sensor used to monitor water quality parameters is the airborne imaging spectrometer for application (AISA), which is a push-broom system that collect spectral-radiance data (upwelling radiance and downwelling solar irradiance) in the visible and NIR range of the electromagnetic spectrum (approximately from 392 to 982 nm with a bandwidth of 7–8 nm). From an altitude of 1,000 m, this sensor has a spatial resolution of 1 m, surpassing the problems caused by medium to low spatial resolutions found in orbital sensors. In [18], AISA imagery was used to estimate chl-a and phycocyanin (PC) concentrations in a mesotrophic reservoir in Central Indiana, USA, based on a series of semi-empirical algorithms. In [19], the authors used the same imagery from the previous study [18] to apply a quasi-analytical algorithm and spatialize the chl-a, backscattering (b_b) and a. In [20], AISA imagery was used to measure chl-a, suspended solids, turbidity and other measures of water clarity from major rivers of Minnesota, USA. Although the use of airborne hyperspectral sensors showed to be an alternative to the development of bio-optical algorithms, because of the expenses of the acquisition and low temporal availability, airborne sensors have not been highly used for water quality monitoring.

An orbital hyperspectral sensor could be the solution for the high costs of flying an airborne sensor, and this was accomplished by the launch of Hyperion, in 2000. However, this sensor was not used in a water quality research because of its signal-to-noise ratio which was very low [21], and also because of its unreliability caused by problems such as radiometric instability. An alternative for the acquisition of hyperspectral images with a medium spatial resolution was the hyperspectral imager for the coastal ocean (HICO), a hyperspectral sensor with 87 spectral bands covering the visible and NIR range (400–900 nm) on-board of the International Space Station (ISS). HICO acquired programmed images from September 2009 to September 2014 with a spatial resolution of 90 m, higher than MERIS (300 m) and MODIS (250, 500 and 1000 m). Since HICO was a sensor developed for the monitoring of aquatic environments, several researches used it to monitor several parameters such as: seagrass and algae mapping [22], cloud removal [23], red tide detection [24], improved chl-a detection [25] and harmful cyanobacteria bloom detection [26]. These studies showed strong relationships between these aquatic constituents and reflectance data which could be used to monitor water quality. They also highlight the importance of having an orbital hyperspectral sensor with a high signal-to-noise ratio to improve the development of bio-optical algorithms for inland and coastal waters. The bio-optical modeling of water quality parameters can be used as a complement to conventional monitoring programs which are usually based on sampling and analyzing of few spots in the aquatic system. Moreover, traditional monitoring programs are costly and time-consuming [27], while bio-optical modeling can quickly provide a synoptic view of the environment.

1.1. Hypothesis

Bio-optical algorithms developed for inland or deep ocean waters are unable to uptake empirical algorithms developed especially for extreme environments.

1.2. Objectives

In this chapter, we attempted to present an overview of the application of bio-optical algo-rithms to monitor water quality parameters as well as to assess chl-*a* and turbidity from a hypersaline tropical lake (Araruama Lagoon) in Rio de Janeiro, Brazil, using HICO imagery and different bio-optical algorithms. A secondary goal of this chapter was to evaluate the performance of bio-optical algorithms for the estimation of chl-*a* and turbidity in an extreme aquatic system such as the Araruama Lagoon.

2. Study site

The Araruama Lagoon is a hypersaline coastal lagoon located in the central coast of Rio de Janeiro State, Southeastern Brazil, between latitudes 22°50'S and 22°57' S and the longitudes 42°00' W and 42°44' W. It is situated in a micro-region called "Região dos Lagos", around 120 km from Rio de Janeiro City (Figure 1a,b). This region is densely populated showing a population density around 268 habitants per square kilometer [28]. The lagoon area encom-passes five municipalities: Araruama, Arraial do Cabo, Cabo Frio, Iguaba Grande, São Pedro da Aldeia and Cabo Frio (see Figure 1c).

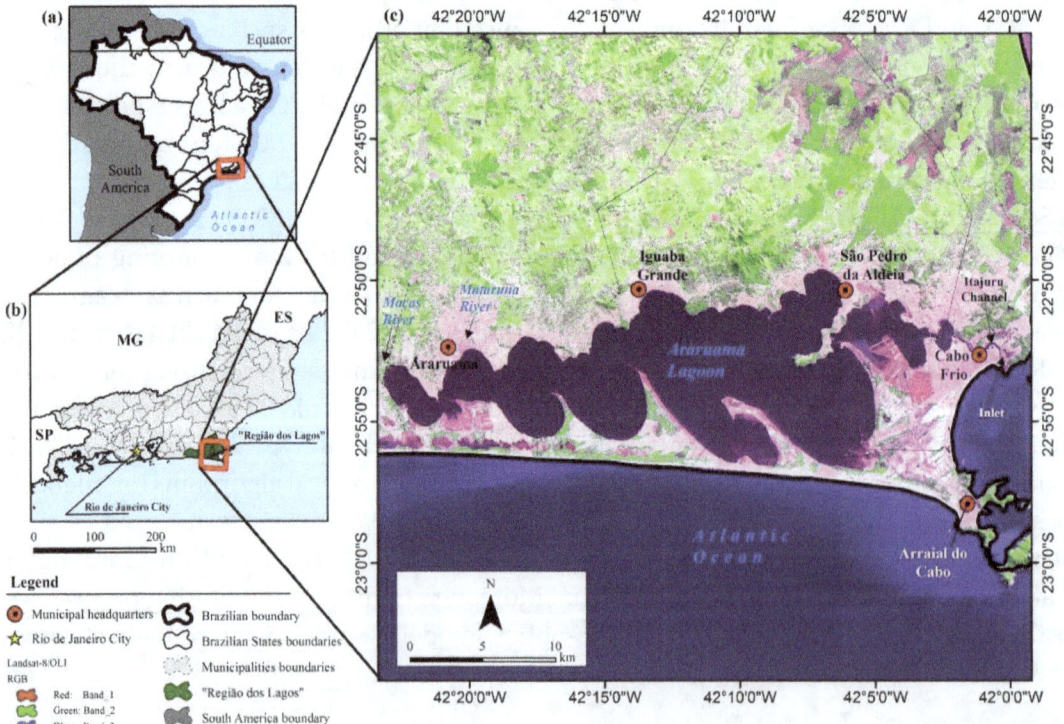

Figure 1. The Araruama Lagoon: (a) Location in Southeastern Brazil, (b) position within the Rio de Janeiro State, and (c) orbital image of the Araruama Lagoon acquired on 1st August 2015 by the Operational Land Imager (OLI) on-board Landsat-8 satellite. The satellite images are presented in false color composition R4G5B2.

From the morphological point view, the Araruama Lagoon consists of a series of elongated spits and shallow embayment presenting a longitudinal elongated shape with around 35 km in length and a mean width of 8 km; the maximum width is around 13 km. The surface area is around 220 km² and the depth ranges from 1 to 17 m; the mean depth is around 3 m [29]. The only connection between the Araruama Lagoon and sea, the Itajuru Channel, is located in the Cabo Frio City, Northeastern portion of the lagoon (see Figure 1c). The drainage basin covers around 320 km², and permanent sources of freshwater come from Moças River and Mataruna River, in the Western portion of the Lagoon (see Figure 1c); the two rivers present a combined discharge of 1 m³/s [30].

The salinity of the Araruama Lagoon ranges from 35 to 43 practical salinity unit (psu) in the Itajuru Channel and from 46 to 56 psu in the main body of the Lagoon, being the salinity mainly balanced by the climatology of the area [29]. According to the Köppen-Ginger classification scheme [31], the climate in the region can be classified as Tropical Monsoon (Am) with rainfall ranging between 36 (August) and 101 mm per month (December) and the air temperature ranging from 21 (August) and 25.4°C (February–March) along the year ([32], see Figure 2); the mean annual precipitation is 771 mm per year and the mean air temperature is around 23°C.

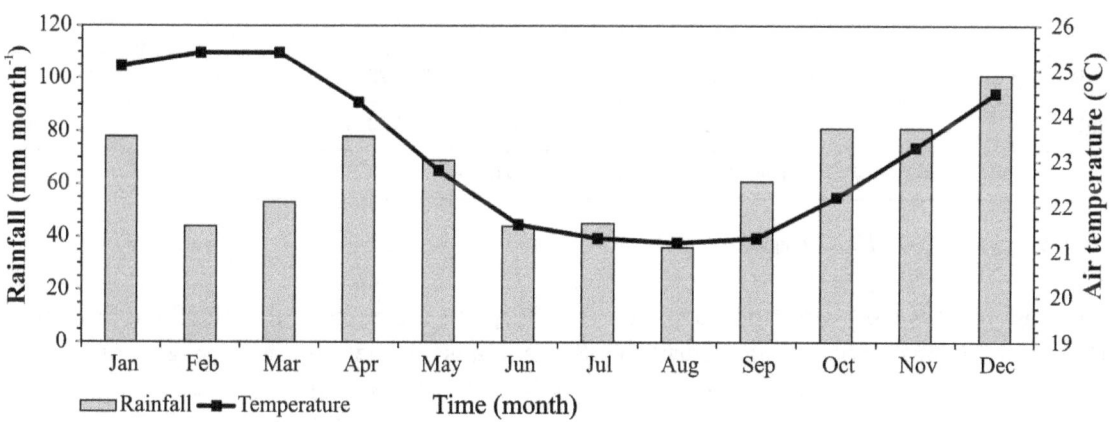

Figure 2. Climatological (1961–1990) monthly rainfall and air temperatures in the Araruama Lagoon region. Data registered on Cabo Frio meteorological station (Lat. -22.98°; Long. 42.03°). Source [32].

The water quality in the Araruama Lagoon has changed over the time, showing an increasing eutrophication along the past few years as a result of the increasing urban growth in the Região dos Lagos [33]. According to the Trophic State Index (TSI) classification scheme proposed by [34], the Araruama Lagoon can be classified as eutrophic environment, with an average total phosphorous concentration around 0.09 mg/L and the average chl-a concentration around 11.7 µg/L [35]. Cyanobacteria (*Synechococcus* sp, *Oscillatoria* sp and *Synechocystis* sp) is the dominant community in the water column along the year (around 84% of the total cell count), followed by Diatomaceous (around 7% of the total cell count) and Dinoflagellates (around 5% of the total cell count) [36].

3. Materials and methods

3.1. Remote sensing data

HICO imageries of Araruama Lagoon were acquired from HICO's website database at Oregon State University (OSU) [37]. The acquisition of the images over Araruama Lagoon occurred from 2011 to 2013, where only images without cloud cover over the lagoon were selected. HICO images are available with a Level 1B of processing, which corresponds to the radiance in the top of the atmosphere (L_{TOA}) given in $Wm^{-2}\mu m^{-1}sr^{-1}$ after the application of a division factor of 50. Table 2 lists the HICO imagery with clear sky over the Araruama Lagoon.

Year	Julian day	Image Name
2011	220	iss.2011220.0808.120519.L1B.GLT_Habitat_Brazil.v04.7594.20110809180058.100m.hico
2012	037	iss.2012037.0206.112051.L1B.GLT_Habitat_Brazil.v04.9365.20120206182247.100m.hico
2012	040	iss.2012040.0209.100728.L1B.GLT_Habitat_Brazil.v04.9394.20120209193848.100m.hico
2012	094	iss.2012094.0403.122511.L1B.GLT_Habitat_Brazil.v04.9907.20120403190851.100m.hico
2012	282	iss.2012282.1008.094232.L1B.GLT_Habitat_Brazil.v04.11631.20121009174522.100m.hico
2013	152	iss.2013152.0601.114032.L1B.GLT_Habitat_Brazil.v04.13707.20130603175752.100m.hico
2013	215	iss.2013215.0803.110724.L1B.GLT_Habitat_Brazil.v04.14303.20130805151206.100m.hico
2013	279	iss.2013279.1006.094546.L1B.GLT_Habitat_Brazil.v04.14826.20131007170614.100m.hico

Table 2. List of clear sky HICO images over Araruama Lagoon

All these images were atmospherically corrected by the Second Signal in the Solar Spectrum (6S) implementation of Tafkaa algorithm [38]. Tafkaa is a radiative transfer algorithm developed mainly for applications in the field of oceanic hyperspectral remote sensing, and it is based on an earlier code named ATmospheric REMoval (ATREM) [39]. Tafkaa is available for processing HICO images online via a web tool [37], with prior registration. For the atmospheric correction over the Araruama Lagoon, the aerosol model was set to "maritime" and the atmospheric model was set to "tropical", since these characteristics seem to be the more appropriate for the study site. The final products of this process are delivered in units of R_{rs}, sr^{-1}, and the spectra from the 12 sampling points of the Inter-Municipal Consortium Lagos São João [35] were obtained for the bio-optical modeling.

3.2. Limnological data

Chl-a ($\mu g/L$) and turbidity (NTU) data were acquired from the reports from the Inter-municipal Consortium Lagos São João, which are available at [35]. This consortium collects monthly data from 12 sampling points in the Araruama Lagoon and has the goal to propose and execute actions to recover the environment in the watershed of three different lagoons (Jaconé,

Saquarema e Araruama) in Rio de Janeiro State, Brazil. Reports matching the HICO imagery (i.e., field campaigns that were carried out on the nearest date as possible as the images acquisition, Table 2) were used to acquire chl-*a* and turbidity data and a total of 87 useful sampling locations were found. These data were divided into two datasets: calibration (53 sampling points using data from 2011 to 2012) and the validation (34 sampling points using data from 2013). Figure 3 shows box-plots to access the statistical distribution of the chl-*a* (μg/L) and turbidity (NTU) values from Araruama Lagoon that were used to calibrate and validate the bio-optical algorithms.

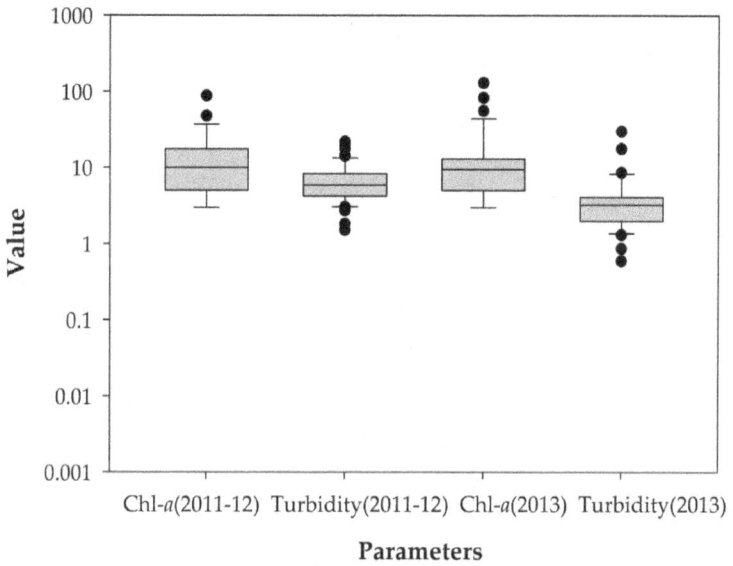

Figure 3. Box-plots of chl-*a* (μg/L) and turbidity (NTU) values used for calibration (2011–12) and for validation (2013) of the bio-optical algorithms.

3.3. Bio-optical algorithms

Several empirical and semi-empirical bio-optical algorithms for chl-*a* and turbidity have been developed in the past decade. Since HICO is a hyperspectral sensor, it is possible to apply several bio-optical algorithms which use different spectral bands. For the estimation of chl-*a* spectral features such low reflectance (troughs) at ~430 nm and ~670 nm caused by the absorption of chl-*a* and a phytoplankton scattering peak at ~700 nm are commonly used in the development of semi-empirical algorithms. The combination of these spectral features makes the ratio of between R_{rs} around 700 and 670 nm [40] widely used for bio-optical algorithm for estimating chl-*a* concentration in turbid waters. There are other algorithms that employ slight variations of this ratio, such as the three band algorithm [41] which uses a third band to minimize the effect of scattering which should be a spectral band with minimal absorption (usually around 750 nm). Another variation is the four band algorithm [42] which includes a spectral band located near 700 nm to enhance the minimization of scattering of suspended matter at the NIR and the absorption by water. Recently, a Normalize Difference Chlorophyll

Index (NDCI) [43] was proposed to estimate chl-*a* concentrations in turbid waters and also used the relationship between 700 and 670 nm. All these algorithms were proposed to estimate chl-*a* concentration in turbid waters; however, for deep ocean waters chl-*a* can also be estimated by algorithms based on band ratios focusing on the chl-*a* absorption around 430 nm and the scattering of particulate matter around 560 nm [44].

Name	Algorithm	Reference
	Chl-a Algorithms	
2BDA	$\dfrac{R_{rs}(band_{54})}{R_{rs}(band_{47})}$	[40]
3BDA	$\dfrac{R_{rs}(band_{62})}{\left[R_{rs}(band_{47}) - R_{rs}(band_{54})\right]}$	[41]
4BDA	$\left\{\left[\dfrac{1}{R_{rs}(band_{46})}\right] - \left[\dfrac{1}{R_{rs}(band_{51})}\right]\right\} \Big/ \left\{\left[\dfrac{1}{R_{rs}(band_{60})}\right] - \left[\dfrac{1}{R_{rs}(band_{54})}\right]\right\}$	[42]
NDCI	$\dfrac{\left[R_{rs}(band_{54}) - R_{rs}(band_{47})\right]}{\left[R_{rs}(band_{54}) + R_{rs}(band_{47})\right]}$	[43]
OC3A	$\dfrac{R_{rs}(band_8)}{R_{rs}(band_{27})}$	[44]
OC3B	$\dfrac{R_{rs}(band_{16})}{R_{rs}(band_{27})}$	[44]
OC3C	$\dfrac{R_{rs}(band_{20})}{R_{rs}(band_{27})}$	[44]
	Turbidity Algorithms	
1BDA	$R_{rs}(band_{43})$	[45]
2BDA	$\dfrac{R_{rs}(band_{81})}{R_{rs}(band_{43})}$	[46]
LSBA	$R_{rs}(band_{15}) + R_{rs}(band_{28})$	[47]

Table 3. List of bio-optical algorithms for chl-*a* and turbidity using HICO spectral bands

Turbidity is usually identified by the high reflectance in the red and NIR spectral bands and is usually correlated to the total suspended solids concentration. Therefore, bio-optical algorithms for TSS can be used to estimate turbidity. The simplest algorithm uses the R_{rs} at 645 nm to estimate turbidity [45]; however, other algorithms were also proposed to estimate turbidity using the relationship between NIR and red spectral bands, such as the band ratio proposed by [46]. Another algorithm to estimate turbidity is based on the sum of R_{rs} in the blue and green spectral bands [47]. However, most of the algorithms were developed for inland, coastal or oceanic waters and have not been applied in extreme environments such as the hypersaline Araruama Lagoon. Table 3 lists the published semi-empirical bio-optical algorithms for chl-*a* and turbidity evaluated in this chapter. The algorithms in this list are expressed according to the 87 HICO's spectral bands.

3.4. Bio-optical algorithm development

Since the Araruama Lagoon is a hypersaline aquatic system, and the bio-optical algorithms listed in the previous section were develop for fresh or oceanic waters, we developed two empirical algorithms for the estimation of chl-*a* and turbidity. The development of these empirical algorithms was conducted by calculating the correlation among different band ratios and the concentrations of chl-*a* and turbidity values. To perform this analysis, we used a web tool named Interactive Correlation Environment (ICE) described by [48] and available at [49]. This web tool builds a two-dimensional correlation plot of the HICO's R_{rs} and its relation to the interested limnological parameter (i.e., chl-*a* or turbidity). The two-dimensional color correlation plot can cover all possible band ratios, which in the HICOs case is equal to 7,569 possible combinations, making it a useful tool for the analysis of hyperspectral measurements with a large number of spectral bands.

3.5. Bio-optical algorithms comparison

As described in Section 3.2, the data were divided in calibration (2011–12, 53 sampling points) and validation (2013, 34 sampling points) datasets. For the calibration dataset, a linear regression analysis was computed by the values of slope and intercept for each of the algorithms listed on Table 3 plus the two empirical algorithms developed by the use of ICE. The determination coefficient (R^2) was also computed and the algorithms that had the highest R^2 values were used for validation.

The validation process was computed by analysing a scatter plot between the measured and the estimated values of chl-*a* and turbidity. For chl-*a*, the concentration values were transformed to log (chl-*a*) and for the turbidity, no transformation was needed. We also used errors estimators such as the root mean squared error (RMSE in µg/L or NTU, equation 1) and the normalized root mean squared error (NRMSE, equation 2) to evaluate the performance of the bio-optical algorithms after their calibration.

$$RMSE = \sqrt{\frac{1}{n}\sum_{i=1}^{n}(y_i - x_i)^2} \tag{1}$$

$$NRMSE = \frac{RMSE}{\left(y_{i,max} - y_{i,min}\right)} \tag{2}$$

where: y_i and x_i are the measured and predicted chl-*a*/turbidity values, respectively. In the *i*-th sample, $y_{i,max}$ and $y_{i,min}$ are the maximum and minimum chl-*a*/turbidity values, respectively.

4. Results and discussions

4.1. ICE's results

To compute the two-dimensional color correlation plot, the R_{rs} spectra were extracted from HICO imagery over the sampling locations in the Araruama Lagoon (Figure 4). Figure 4 presents the R_{rs} spectra of the calibration (Figure 4A) and validation (Figure 4B) datasets. Both datasets presented spectral features of an eutrophic water, with a high reflectance peak in the green range around 550 nm, a trough near 620 nm, another trough around 665 nm and a peak in the NIR around 700 nm. The reflectance peak around 550 nm represents the minimal absorption of all algal pigments and the scattering of non-organic suspended matter and phytoplankton cell walls [50]. The trough around 620 nm occurs due to the absorption of phycocyanin, a phycobillin presented in inland water cyanobacteria [51–52]. The trough around 665 nm is due to the absorption of chl-a in the red range of the spectrum and the peak around 700 nm is also dependable of the chl-a since it represents the scattering of the suspended matter which includes algal biomass [53]. The two-dimensional color correlation plot was computed using the R_{rs} from the calibration dataset as well as the limnological dataset presented in Section 3.2.

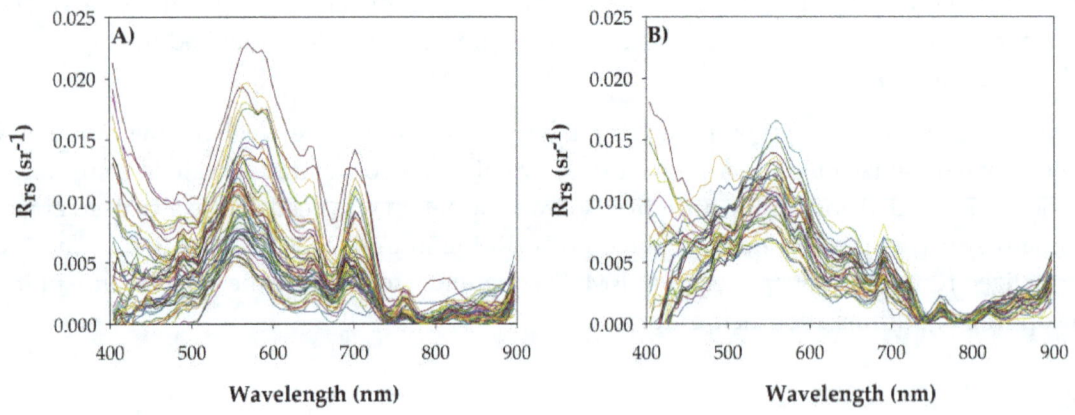

Figure 4. R_{rs} spectra from HICO imagery after atmospheric correction. A) R_{rs} spectra from 2011 and 2012 used for the calibration; B) R_{rs} spectra from 2013 used for the validation.

The use of ICE generates two different two-dimensional color correlation plots, one for chl-a (Figure 5A) and one for turbidity (Figure 5C). Using the filtering tool implemented in ICE, it was possible to select the band ratio that gives the highest R^2 for each of the parameters. Figure 5B shows the filtered plot for chl-a estimation which highlights only the band ratios with high R^2, Figure 5D shows the same filtered plot for the turbidity. Since the choices of spectral bands are only based on the statistical estimators among all possible band ratios, the algorithms derived from this web tool can be classified as empirical, and does not have a biophysical background to support the spectral bands used in each band ratio.

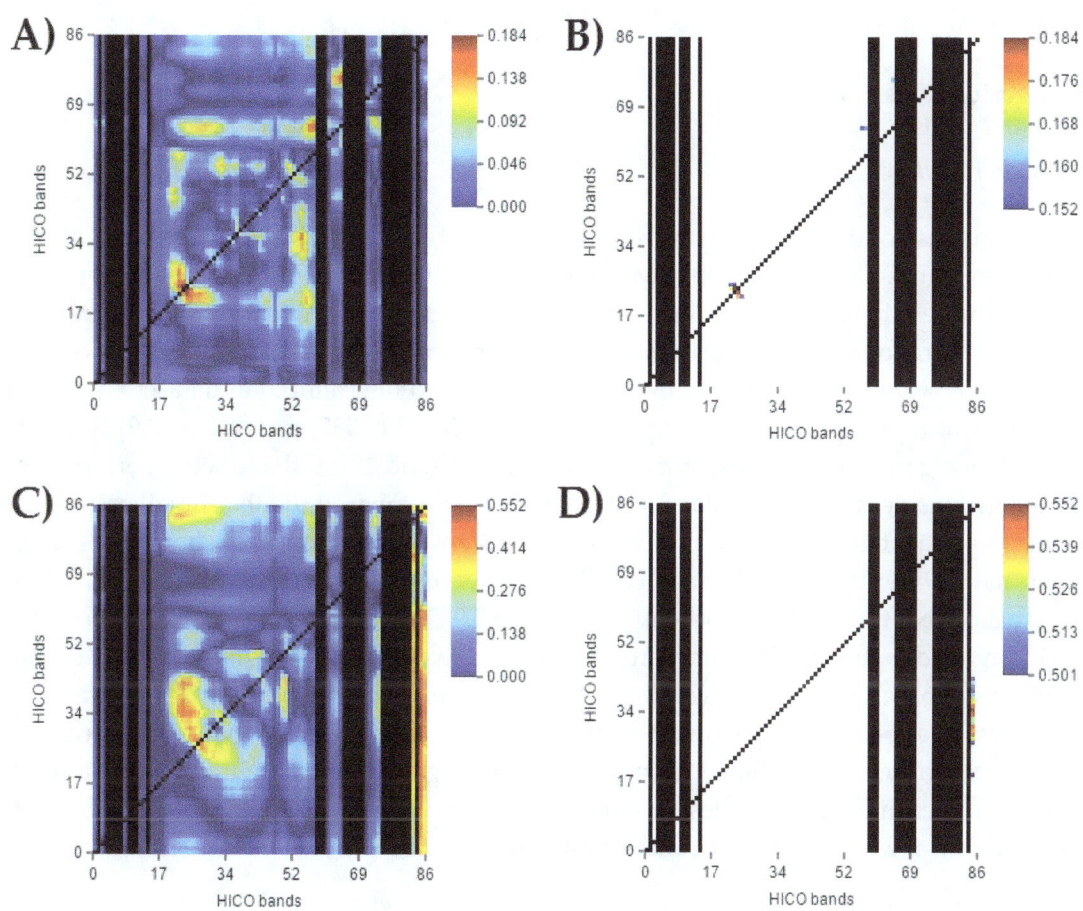

Figure 5. Two-dimensional color correlation plots produced by the web tool. (A) For chl-*a* estimation; (B) After the filtering the chl-*a* plot; (C) For turbidity estimation; (D) After the filtering the turbidity plot.

For chl-*a* the best R^2 was found in correlation to the ratio between band 26 and band 25 which in wavelengths are around 547 and 541 nm, respectively. This relationship is totally empirical and since both bands are very close to each other, the value from this ratio is probably close to 1. For the turbidity band ratio, ICE selected the ratio between band 36 and band 87, which respectively corresponds to 604 and 896 nm. Although it is an empirical model, these two bands can be justified by the fact in both wavelengths the suspended matter will have a high scattering, and if the suspended matter is high, the turbidity will also be high. The formulation and name of these two empirical bio-optical algorithms were described in Table 4.

Name	Parameter	Algorithm
EMPC	Chl-a	$\dfrac{R_{rs}(band_{26})}{R_{rs}(band_{25})}$
EMPT	Turbidity	$\dfrac{R_{rs}(band_{36})}{R_{rs}(band_{87})}$

Table 4. Empirical algorithms for chl-a and turbidity developed using ICE

4.2. Algorithms performances

4.2.1. Calibration

Calibration was conducted using the semi-empirical (Table 3) and empirical (Table 4) bio-optical algorithms. Linear regressions were computed between bio-optical algorithms and chl-a and TSS values; the R^2, slope and intercept from each of the regressions were shown in Table 5. For the calibration of chl-a algorithms, all the algorithms showed a poor performance with the highest R^2 value of 0.087 found by applying EMPC to the calibration dataset. The other high values of R^2 were found by using OC3C (0.065), OC3B (0.037) and 4BDA (0.011), which showed that algorithms developed for deep ocean (OC3B and OC3C) have better performance than algorithms developed for inland waters (4BDA). If compared to other Brazilian tropical inland water aquatic systems, the performance of 2BDA, 3BDA and NDCI showed R^2 values higher than 0.9 during the calibration step [12]. However, for the Araruama Lagoon, the R^2 values from these three algorithms were lower than 0.003. This difference in the performance could be related to the fact that Araruama Lagoon is a hypersaline environment and the high concentration of salt in the water could be masking the results, although the R_{rs} spectra show the features of a reservoir dominated by cyanobacteria. This poor performance in the calibration of all algorithms highlights the importance of having an extreme environment as a study site in bio-optical modeling studies, since one of the goals of this field is to have an algorithm that can perform well in different aquatic systems.

The poor performance of all algorithms could be associated to the fact that none of these algorithms were developed for hypersaline aquatic systems, which make their calibration difficult in this type of environment. Another source of error could be associated to the temporal window between the image acquisition and field sampling. Since we are using ground truth data that are collected as part of a routine monthly monitoring, we could not find an exact match with temporal windows ranging from 2 to more than 10 days. This can lead to erroneous interpretations since the dynamics of parameters, mainly the biotic ones such as phytoplankton, in the water column can change within days according to the environment dynamics. Adopting a 3-days window, the calibrations showed in Table 5 improved mainly for the chl-a algorithms. Table 6 shows the calibrations using only the images within the 3-days window from the field campaign, which shows that EMPC got a R^2 value of 0.43, while using the entire dataset the R^2 value was around 0.08. For the turbidity estimations, the improvement was not big as for the chl-a estimations varying from an R^2 of 0.574 for the 1BDA using the entire dataset to an R^2 of 0.596 using only the 3-days window data. These results

Chl-a algorithms			
Name	R²	Slope	Intercept
2BDA	<0.001	-0.026	14.369
3BDA	0.002	-0.256	14.270
4BDA	0.011	9.806	14.097
NDCI	0.003	7.878	12.561
OC3A	0.006	-4.510	15.753
OC3B	0.037	-16.635	21.911
OC3C	0.065	-29.037	28.519
EMPC	0.087	101.760	-98.513
Turbidity algorithms			
Name	R²	Slope	Intercept
1BDA	0.574	1145.1	1.1929
2BDA	0.127	-3.5418	8.2162
LSBA	0.385	453.18	0.3646
EMPT	0.450	1.2385	3.518

Table 5. R^2, slope and intercept of the linear regression from the bio-optical algorithms tested (shaded areas represents the algorithms that were used for validation)

showed that the calibration is most affected by the temporal window when a biotic parameter is being analysed.

Chl-a algorithms			
Name	R²	Slope	Intercept
4BDA	0.049	19.022	15.061
OC3A	0.223	-33.516	29.576
OC3B	0.186	-23.12	23.798
EMPC	0.430	159.3	-164.79
Turbidity algorithms			
Name	R²	Slope	Intercept
1BDA	0.596	1031.2	3.054
2BDA	0.211	-3.193	10.512
LSBA	0.509	437.33	2.757
EMPT	0.304	0.890	5.550

Table 6. R^2, slope and intercept of the linear regression from the bio-optical algorithms tested using a 3 days temporal window

If compared to the performance of 2BDA and 3BDA for the estimation of chl-*a* in Taganrog Bay [25] the calibration for Araruama is poor, since the R^2 values found by [25] were 0.84 and 0.87 for the 2DBA and 3BDA, respectively. However, if compared to another hypersaline environment, such as Mono Lake, CA, USA, the results are equivalent since an R^2 of 0.49 was found to correlate chl-*a* and a bio-optical algorithm developed for the airborne visible/infrared imaging spectrometer (AVIRIS) sensor [54]. These results highlight the current need for a global database of bio-optical modeling data for inland waters, as well as the development of more semi- and quasi-analytical algorithms. Although few attempts have been proposed to create a global database of bio-optical modeling data for inland waters [55–56], we still need an organization such as the International Ocean Colour Coordinating Group (IOCCG) to coordinate the protocols, products and database of algorithms and bio-optical modeling data. Thus, the global bio-optical modeling of inland waters will only be achieved when we have algorithms developed and tested for all different types of environments.

4.2.2. Validation

The validation of the bio-optical algorithms with the best R^2 in the calibration was computed using two different methods: validation plots between the measured and the estimated values of chl-*a* and turbidity and error estimators. Figure 6 showed the validation plots for the four bio-optical algorithms analysed in this chapter: 4BDA (Figure 6A), OC3C (Figure 6B), OC3B (Figure 6C) and EMPC (Figure 6D). The dashed red line represents the 1:1 line where the points of the scatter plot should be over that line. In Figure 6A, we observed that the points are vertically distributed showing that there is no variation in the estimated values of log (chl-*a*); however, the error estimator showed that 4BDA has the lowest NRMSE of 24.13% among the tested algorithms. This fact showed that error estimators are only statistical and do not represent well the reality of the distribution of the data. Figure 6B and 6C showed the results for the algorithms proposed for ocean color remote sensing, OC3B and OC3C, the NRMSE for these bio-optical algorithms were 27.37% and 30.80%, respectively. The validation plots for both bio-optical algorithms showed a better distribution than the 4BDA since they showed a better distribution over the 1:1 line. However, both of the ocean color algorithms showed to underestimate (Figures 6B and 6C) the high values of log (chl-*a*). Figure 6D showed the scatter plot for the empirical algorithm (EMPC), which showed a NRMSE of 30.46%. This validation plot showed a similar pattern to the previous semi-empirical models and also underestimate the values of log (chl-*a*).

Based on these results for the chl-*a* bio-optical algorithms, we observed that although the lowest NRMSE was found in the 4BDA, the validation plots showed that the other algorithms showed a better distribution and can explain better the estimation of the log (chl-*a*). Therefore, OC3B showed the best performance with a NRMSE of 27.37% and a validation plot that underestimates the high values of chl-*a*. This underestimation was expected since for the ocean color bio-optical modeling the concentration of chl-*a* is not high and in the case of Araruama Lagoon the concentration can reach 130 µg/L of chl-*a*. As discussed in the calibration results, if compared to another tropical inland water body, the NRMSE are higher. In [12] the NRMSE founded for 2BDA, 3BDA and NDCI were 18.32%, 19.68% and 17.85%, respectively, for bio-optical models calculated using proximal hyperspectral sensor. These differences can be related to the atmospheric correction, which is not needed for the proximal hyperspectral sensor [12], but also for the fact that Araruama Lagoon has a very unique biogeochemical cycling, which could lead to different composition of the water column. Overall, more studies should be conducted in Araruama Lagoon to better understand the optical properties in this aquatic system.

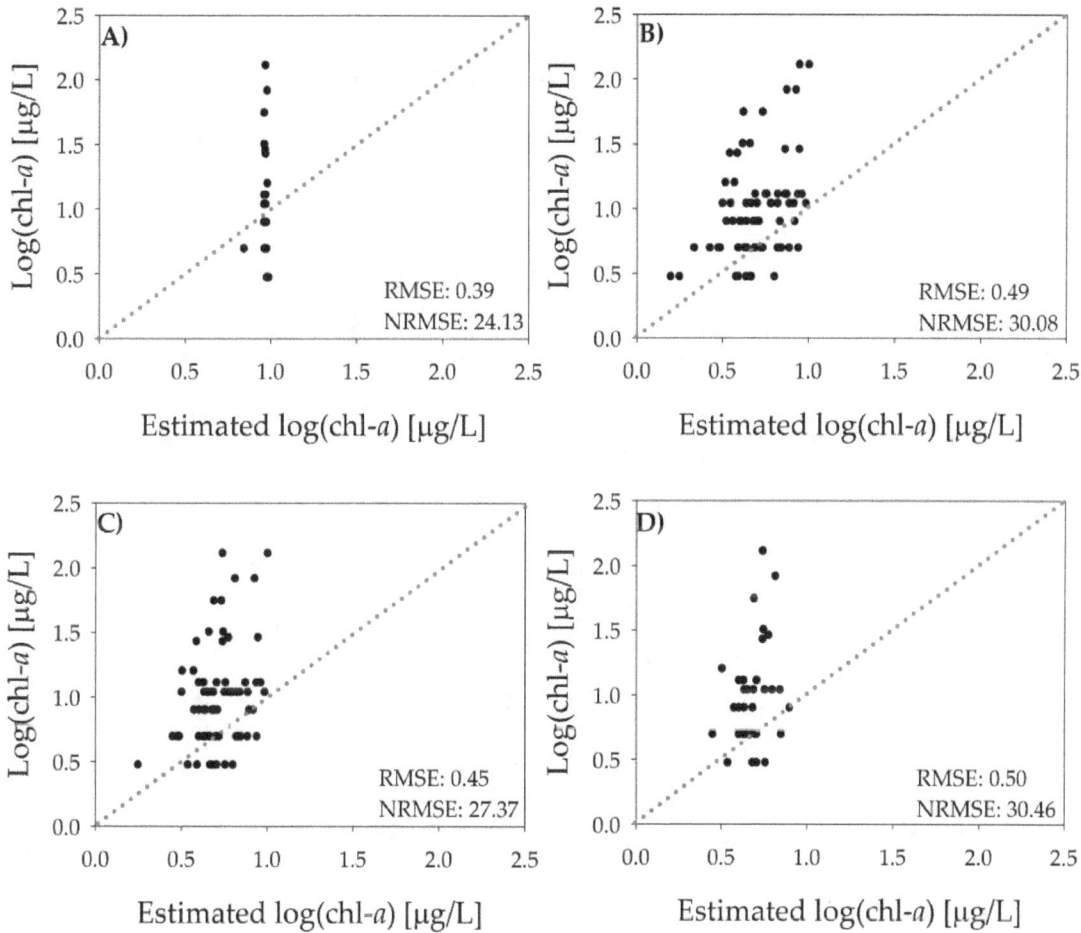

Figure 6. Validation plots for the chl-*a* bio-optical algorithms: (A) 4BDA; (B) OC3C; (C) OC3B; (D) EMPC

The validation for the turbidity bio-optical algorithms showed a better agreement between the validation plots and the error estimators. Figure 7 showed the validation plots for the four bio-optical algorithms analysed in this chapter: 1BDA (Figure 7A), 2BDA (Figure 7B), LSBA (Figure 7C) and EMPT (Figure 7D). The lowest NRMSE was 15.49% and was achieved by applying the 1BDA to the R_{rs} data (Figure 7A); the validation plot also showed most of the points close to the 1:1 line, but as well as for the chl-*a* algorithms for high values of turbidity the algorithms underestimated the values. The empirical algorithm (EMPT) had the second best NRMSE

(17.87%) among the turbidity bio-optical algorithms analysed and the validation plot showed to be similar to the previous algorithm. 2BDA and LSBA showed a NRMSE of 22.37% and 28.04%, respectively, which are higher than the 1BDA and EMPT algorithms. The validation plot for 2BDA and LSBA also showed a worst distribution of the scatter points and also showed an underestimation of the high values of turbidity. Overall, the performance of 1BDA showed the best validation plot and NRMSE value among the four bio-optical algorithms analysed in this chapter.

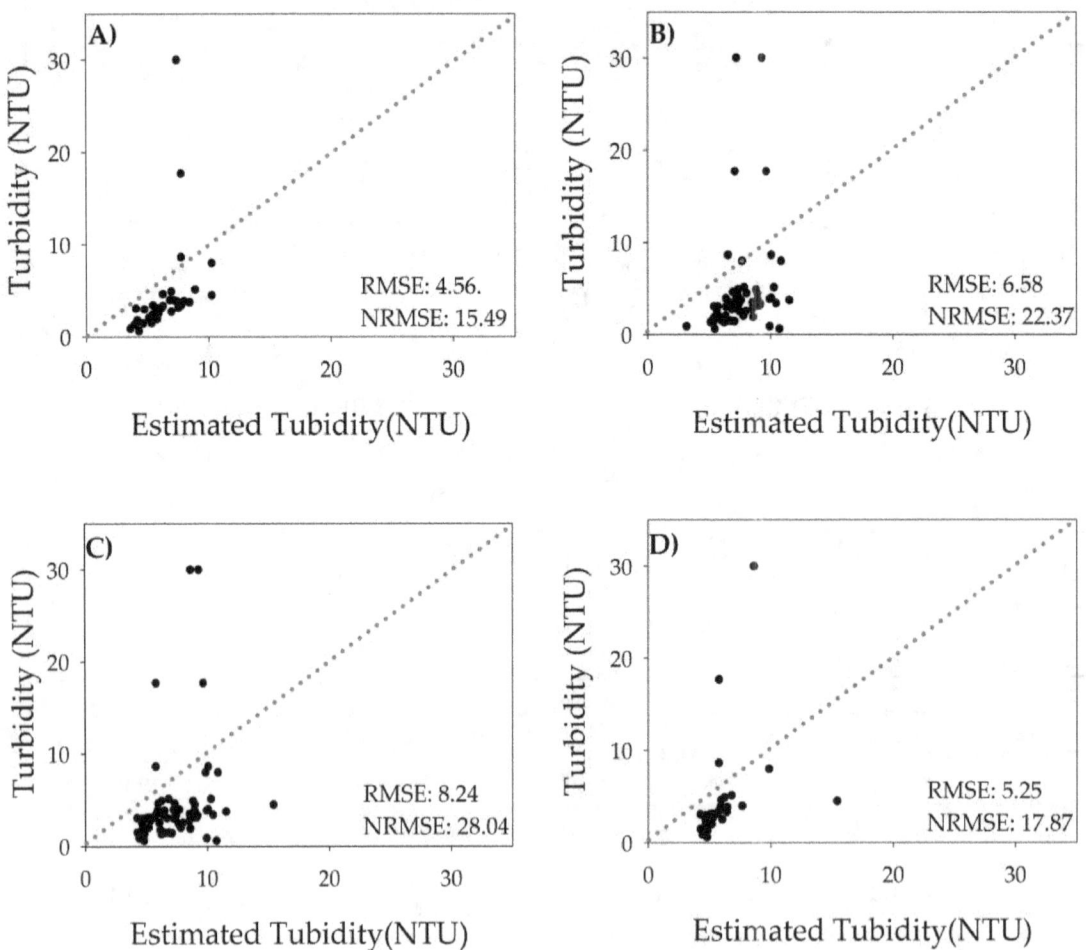

Figure 7. Validation plots for the turbidity bio-optical algorithms: (A) 1BDA; (B) 2BDA; (C) LSBA; (D) EMPT

4.3. Spatial distribution

Applications of bio-optical modeling to monitor water quality in inland waters have been increasing in the past decade, and this increase is also noticed in the public and private sector investments on remote sensing technologies to monitor water quality and quantity. The advantages of using remote sensing technologies over traditional methods to monitor water quality parameters were already discussed in the introduction of this chapter; however, another advantage of using remote sensing is in the spatial distribution of the data. While using traditional methods of water quality monitoring computes the spatial assessment of the water quality by performing spatial interpolations or by geostatistical methods of few sampling points, remote sensing images can provide different values for each pixel within the aquatic system. The difference is that the few sampling points used to interpolate the data for the aquatic system area is now replaced for several pixels values in the image, where the interpolation is not needed; therefore, it does not have the error caused by data interpolation methods. Figure 8 shows the spatial distribution of chl-*a* and turbidity in the Araruama Lagoon where we observed that the west part of the lagoon has the highest values of chl-*a* and for the turbidity regions close to the bays have highest turbidity values. These spatial patterns are related to the hydrodynamic of the aquatic system, and the combination of bio-optical and hydrodynamic modeling [57] is a powerful tool to understand the spatial dynamics of the environment.

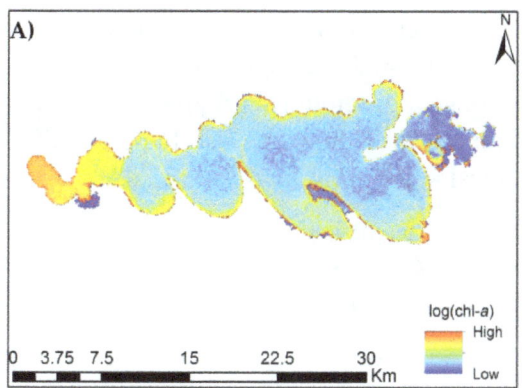 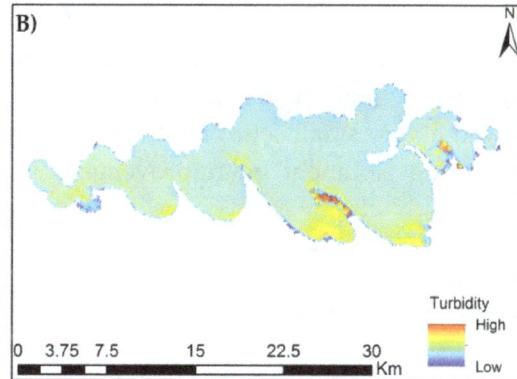

Figure 8. Application of the bio-optical algorithms to the HICO image from Araruama Lagoon acquired on August 3, 2013. (A) Application of calibrated OC3C; (B) Application of calibrated 1BDA

5. Final considerations

Based on the case study of Araruama Lagoon, we observe the need for calibration and validation of bio-optical algorithms in different inland waters since the variability of water column constituents from region to region is big. We also observe that the use of orbital hyperspectral sensors is important for the development of bio-optical modeling due to the number of spectral bands which allow us to study small features, such as the absorption peak of PC around 620 nm. Thus, narrow spectral bands can highlight specific absorption features which can be used in the development and improvement of bio-optical algorithms, mainly the semi- and quasi-analytical algorithms which are based on the radiative transfer theory. Therefore, future hyperspectral missions such as the Hyperspectral Imager SUIte (HISUI), the PRecursore IperSpettrale della Missione Applicativa (PRISMA), and the Environmental

Mapping and Analysis Program (EnMAP) are important for the development of bio-optical modeling.

Moreover, these new hyperspectral missions will support a global mapping of inland water quality which is only possible through multispectral sensors such as Landsat, MODIS and MERIS. However, not all water quality parameters are possible to be measured only using multispectral sensors, for example, the Landsat series which have a poor spectral resolution that does not detect the spectral features such as peaks and trough of chl-*a*. MODIS, on the other hand, have a narrow band moderate spectral resolution; however, its spatial resolution makes the monitoring of small and medium inland water bodies difficult. Therefore, the development of new global hyperspectral sensors will make the assessment of water quality through remote sensing possible because of the high spectral and spatial resolution.

Finally, our case study showed that even by developing an empirical algorithm, the semi-empirical algorithms outperform them. The best performance for chl-*a* bio-optical algorithms was found by applying OC3B (NRMSE of 27.37%) and for turbidity, the 1BDA showed the best performance with a NRMSE of 15.49%. Although the lower errors estimators validation plots (Figures 6 and 7) showed that all algorithms underestimated the high values of chl-a and turbidity, highlighting the need of different calibrations for different water types. This chapter showed a very small set of methods used in bio-optical modeling and also highlighted the need for development and improvement of bio-optical algorithms.

Acknowledgements

We thank the HICO team at Oregon State University (OSU), especially Jasmine Nahorniak for providing access to the database and to all her attention to us. C. A. S. Araújo thanks the Brazilian National Counsel of Technological and Scientific Development (CNPq) for the PCI fellowship (under the grant 300177/2015-1). M. P. Curtarelli also thanks the CNPq for the graduate scholarship (under the grant 161233/2013-9).

Author details

Igor Ogashawara[1*], Marcelo P. Curtarelli[2], Carlos A. S. Araujo[2] and José L. Stech[2]

*Address all correspondence to: igoroga@gmail.com

1 Department of Earth Sciences, Indiana University – Purdue University at Indianapolis (IUPUI), Indianapolis, IN, USA

2 Remote Sensing Division, National Institute for Space Research (INPE), São José dos Campos, SP, Brazil

References

[1] Jensen JR. Remote Sensing of the Environment: An Earth Resource Perspective. 2nd ed. Upper Saddle River, NJ: Prentice-Hall; 2007. 592 p.

[2] Mobley C. Radiative Transfer in the Ocean. In: Steele JH. (Ed.) Encyclopedia of Ocean Sciences. 1st ed. London: Academic Press; 2001. pp. 2321–2330.

[3] Platt T, Hoepffner N, Stuart V, Brown C. (Eds.) Ocean colour? The societal benefits of ocean-colour technology. Reports of the International Ocean-Colour Coordinating Group, No. 7.. 1st ed. Dartmouth, Canada: IOCCG; 2008.

[4] Cox C, Munk W. Measurement of the roughness of the sea surface from photographs of the sun's glitter. J Optic Soc Am 1954;44(11):838–50.

[5] Petzold TJ. Volume scattering functions for selected ocean waters. Ref. 72–28. San Diego, CA: Scripps Institute of Oceanography, University of California; 1972. 79 p.

[6] Jerlov NG. Optical oceanography. Elsevier Oceanographic Series, v. 5. Amsterdam: Elsevier; 1968. 194 p.

[7] Jerlov NG. Marine optics. Elsevier Oceanographic Series, v. 14. ed. Amsterdam: Elsevier; 1976. 231 p.

[8] Preisendorfer RW. Hydrologic optics. Vol. II: Foundations ed. Washington, DC: U.S. Dept. of Commerce; 1976.

[9] Gordon HR, Brown OB, Jacobs MM. Computed Relationships between the Inherent and Apparent Optical Properties of a Flat Homogeneous Ocean. Appl Optic 1975;14(2):417–27.

[10] Bukata RP, Jerome JH, Bruton JE, Jain SC. Determination of inherent optical properties of Lake Ontario coastal waters. Appl Optic 1979;18(23):3926–32.

[11] Ogashawara I. Terminology and classification of bio-optical algorithms. Remote Sens Lett 2015;6(8):613–7. DOI: 10.1080/2150704X.2015.1066523

[12] Augusto-Silva PB, Ogashawara I, Barbosa CCF, Carvalho LAS, Jorge DSF, Fornari CI, Stech JL. Analysis of MERIS reflectance algorithms for estimating chlorophyll-a concentration in a Brazilian reservoir. Remote Sens 2014;6:11689–707. DOI: 10.3390/rs61211689

[13] Matthews MW, Odermatt D. Improved algorithm for routine monitoring of cyanobacteria and eutrophication in inland and near-coastal waters. Remote Sens Environ 2015;156:374–82. DOI: 10.1016/j.rse.2014.10.010

[14] Ogashawara I, Alcantara EH, Curtarelli MP, Adami M, Nascimento RFF, Souza, AF, Stech JL, Kampel M. Performance analysis of MODIS 500-m spatial resolution products for estimating chlorophyll-a concentrations in oligo- to meso-trophic waters case

study: Itumbiara Reservoir, Brazil. Remote Sens 2014;6:1634–53. DOI: 10.3390/rs6021634

[15] Feng L, Hu C, Chen X, Qingjun S. Influence of the Three Gorges Dam on total suspended matters in the Yangtze Estuary and its adjacent coastal waters: Observations from MODIS. Remote Sens Environ 2014;140:779–88. DOI: 10.1016/j.rse.2013.10.002

[16] El-Alem A, Chokmani K, Laurion I, El-Adlouni SE. Comparative analysis of four models to estimate chlorophyll-a concentration in case-2 waters using MODerate resolution imaging spectroradiometer (MODIS) imagery. Remote Sens 2012;4:2373–400. DOI: 10.3390/rs4082373

[17] Hu C. A novel ocean color index to detect floating algae in the global oceans. Remote Sens Environ 2009;113:2118–29. DOI: 10.1016/j.rse.2009.05.012

[18] Li L, Sengpiel RE, Pascual DL, Tedesco LP, Wilson JS, Soyeux E. Using hyperspectral remote sensing to estimate chlorophyll-a and phycocyanin in a mesotrophic reservoir. Int J Remote Sens 2010;31(15):4147–62. DOI: 10.1080/01431161003789549

[19] Li L, Li L, Song K, Li Y, Tedesco LP, Shi K, Li Z. An inversion model for deriving inherent optical properties of inland waters: Establishment, validation and application. Remote Sens Environ 2013;135:150–66. DOI: 10.1016/j.rse.2013.03.031

[20] Olmanson LG, Brezonik PL, Bauer ME. Airborne hyperspectral remote sensing to assess spatial distribution of water quality characteristics in large rivers: The Mississippi River and its tributaries in Minnesota. Remote Sens Environ 2013;130:254–65. DOI: 10.1016/j.rse.2012.11.023

[21] Hu C, Feng L, Lee L, Davis CO, Mannino A, McClain CR, Franz BA. Dynamic range and sensitivity requirements of satellite ocean color sensors: Learning from the past. Appl Optic 2012;51(25):6045–62. DOI: 10.1364/AO.51.006045

[22] Cho HJ, Ogashawara I, Mishra D, White J, Kamerosky A, Morris L, Clarker C, Simpson A, Banisakher D. Evaluating Hyperspectral Imager for the Coastal Ocean (HICO) data for seagrass mapping in Indian River Lagoon, FL. GISci Remote Sens 2014;51:120–38. DOI: 10.1080/15481603.2014.895577

[23] Gao BC, Li RR. Removal of think cirrus scattering effects for remote sensing of ocean color from space. IEEE Geosci Remote Sens Lett 2012;9:972–6. DOI: 10.1109/LGRS.2012.2187876

[24] Ryan JP, Davis CO, Tufllaro NB, Kudela RM, Gao BC. Application of the hyperspectral imager for the coastal ocean to phytoplankton ecology studies in Monterey Bay, CA USA. Remote Sens 2014;6:1007–25. DOI: 10.3390/rs6021007

[25] Moses WJ, Gitelson AA, Berdnikov S, Bowles JH, Povazhnyi V, Sapryngin V, Wagner EJ, Patterson KW. HICO-based NIR-red models for estimating chlorophyll-a concentration in productive coastal waters. IEEE Geosci Remote Sens Lett 2014;11(6):1111–5. DOI: 10.1109/LGRS.2013.2287458

[26] Kudela RM, Palacios SL, Austerberry DC, Accorsi EK, Guild LS, Torres-Perez JL. Application of hyperspectral remote sensing to cyanobacterial blooms in inland waters. Remote Sens Environ 2015;167:196–205. DOI: 10.1016/j.rse.2015.01.025

[27] Duan H, Ma R, Xu J, Zhang Y, Zhang B. Comparison of different semi-empirical algorithms to estimate chlorophyll-a concentration in inland lake water. Environ Monitor Assess 2010;170:231–44. DOI: 10.1007/s10661-009-1228-7

[28] Instituto Brasileiro de Geografia e Estatística (IBGE). Demographic Census [Internet]. 2010. Available from: http://www.ibge.gov.br/home/estatistica/populacao/censo2010/ [Accessed: 01 Sep 2015]

[29] Kjerfve B, Schettini CAF, Knoppers B, Lessa G, Ferreira HO. Hydrology and salt balance in a large, hypersaline coastal lagoon: Lagoa de Araruama, Brazil. Estuarine, Coastal Shelf Sci 1996;42:701–25.

[30] Braga CZF, Vianna ML, Kjerfve B. Environmental characterization of a hypersaline coastal lagoon from Landsat-5 Thematic Mapper data. Int J Remote Sens 2003;24:3219–34. DOI: 10.1080/0143116031000075099

[31] Peel MC, Finlayson BL, McMahon TA. Updated world map of the Köppen-Geiger climate classification. Hydrol Earth System Sci 2007;11:1633–44. DOI: 10.5194/hess-11-1633-2007

[32] Empresa Brasileira de Pesquisa Agropecuária (EMBRAPA). Brazilian climatic database [Internet]. 2003. Available from: http://www.bdclima.cnpm.embrapa.br/ [Accessed: 01 Sep 2015]

[33] Mello TBM. Caracterização Biogeoquímica da Lagoa de Araruama, RJ [thesis]. Niterói: Universidade Federal Fluminense - Instituto de Química; 2007. 82 p. [in Portuguese]

[34] Lamparelli MC. Grau de trofia em corpos d'água do estado de São Paulo: avaliação dos métodos de monitoramento [dissertation]. São Paulo: São Paulo University - Department of Ecology; 2004. 235 p. [in Portuguese]

[35] Comitê de Bacias Lagos São João. Phytoplankton monitoring of the Araruama Lake [Internet]. 2012. Available from: http://www.lagossaojoao.org.br/nc-relatorioqualiaguas.htm [Accessed: 25 Aug 2015] [in Portuguese]

[36] Comitê de Bacias Lagos São João. Monthly water quality Report [Internet]. 2014. Available from: http://www.lagossaojoao.org.br/nc-relatorioqualiaguas.htm [Accessed: 25 Aug 2015] [in Portuguese]

[37] Oregon State University. HICO - Hyperspectral Imager for the Coastal Ocean [Internet]. 2015. Available from: http://hico.coas.oregonstate.edu/ [Accessed: 01 Sep 2015]

[38] Gao BC, Montes MJ, Ahmad Z, Davis CO. Atmospheric correction algorithm for hyperspectral remote sensing of ocean color from space. Appl Optic 2000;39(6):887–96. DOI: 10.1364/AO.39.000887

[39] Gao BC, Goetz AFH. Column atmospheric water vapor and vegetation liquid water retrievals from airborne imaging spectrometer data. J Geophys Res 1990;95(D4):3549–64. DOI: 10.1029/JD095iD04p03549

[40] Dall'Olmo G, Gitelson AA. Effect of bio-optical parameter variability on the remote estimation of chlorophyll-a concentration in turbid productive waters: Experimental results. Appl Optic 2005;44:412–22. DOI: 10.1364/AO.44.000412

[41] Gitelson AA, Gritz U, Merzlyak MN. Relationships between leaf chlorophyll content and spectral reflectance and algorithms for non-destructive chlorophyll assessment in higher plant leaves. J Plant Physiol 2003;160:271–82. DOI: 10.1078/0176-1617-00887

[42] Le C, Li Y, Zha Y, Sun D, Huang C, Lu H. A four-band semi-analytical model for estimating chlorophyll a in highly turbid lakes: The case of Taihu Lake, China. Remote Sens Environ 2009;113:1175–82. DOI: 10.1016/j.rse.2009.02.005

[43] Mishra S, Mishra DR. Normalized difference chlorophyll index: A novel model for remote estimation of chlorophyll-a concentration in turbid productive waters. Remote Sens Environ 2012;117:394–406. DOI: 10.1016/j.rse.2011.10.016

[44] NASA Ocean Biology Processing Group. Ocean Color Chlorophyll (OC) v6 [Internet]. 18 Mar 2010. Available from: http://oceancolor.gsfc.nasa.gov/REPROCESSING/R2009/ocv6/ [Accessed: 02 Sep 2015]

[45] Doxaran D, Babin DM, Leymarie E. Near-infrared light scattering by particles in coastal waters. Optics Exp 2007;15:12834–49. DOI: 10.1364/OE.15.012834

[46] Doxaran D, Froidefond JM, Castaing P, Babin M. Dynamics of the turbidity maximum zone in a macrotidal estuary (the Gironde, France): Observations from field and MODIS satellite data. Estuar Coastal Shelf Sci 2009;81(3):321–32. DOI: 10.1016/j.ecss.2008.11.013

[47] Zhang Y, Lin S, Liu J, Qian X, Ge Y. Time-series MODIS image-based retrieval and distribution analysis of total suspended matter concentrations in Lake Taihu (China). Int J Environ Res Public Health 2010;7(9):3545–60. DOI: 10.3390/ijerph7093545

[48] Ogashawara I, Curtarelli MP, Souza AF, Augusto-Silva PB, Alcântara EH, Stech JL. Interactive correlation environment (ICE) — a statistical web tool for data collinearity analysis. Remote Sens 2014;6:3059–74. DOI: 10.3390/rs6043059

[49] Hidrosfera INPE. ICE - Interactive Correlation Environment [Internet]. 2014. Available from: http://www.dsr.inpe.br/hidrosfera/ice/ [Accessed: 02 Sep 2015]

[50] Gitelson AA, Yacobi YZ, Schalles JF, Rundquist DC, Han L, Stark R, Etzion D. Remote estimation of phytoplankton density in productive waters. Arch Hydrobiol 2000;55:121–36.

[51] Ogashawara I, Mishra DR, Mishra S, Curtarelli MP, Stech JL. A performance review of reflectance based algorithms for predicting phycocyanin concentrations in inland waters. Remote Sens 2013;5:4774–98. DOI: 10.3390/rs5104774

[52] Dekker AG. Detection of Optical Water Quality Parameters for Eutrophic Waters by High [dissertation]. Amsterdam, the Netherlands: Vrije Universiteit; 1993.

[53] Gitelson AA. The peak near 700 nm on radiance spectra of algae and water: relationships of its magnitude and position with chlorophyll concentration. Int J Remote Sens 1992;13(17):3367–73. DOI: 10.1080/01431169208904125

[54] Melack JM, Gastil M. Airborne remote sensing of chlorophyll distributions in Mono Lake, California. Hydrobiologia 2001;466:31–8.

[55] GloboLakes. GloboLakes - Global Observatory of Lake Responses to Environmental Change [Internet]. [Updated: 2014]. Available from: http://www.globolakes.ac.uk/ [Accessed: 02 Sep 2015]

[56] GEO. Group on Earth Observations [Internet]. [Updated: 2015]. Available from: https://www.earthobservations.org/index.php [Accessed: 02 Sep 2015]

[57] Curtarelli MP, Ogashawara I, Alcântara EH, Stech JL. Coupling remote sensing bio-optical and three-dimensional hydrodynamic modeling to study the phytoplankton dynamics in a tropical hydroelectric reservoir. Remote Sens Environ 2015;157:185–98. DOI: 10.1016/j.rse.2014.06.013

6

Remote Sensing for Natural or Man-made Disasters and Environmental Changes

Monika Gähler

Additional information is available at the end of the chapter

Abstract

Disasters can cause drastic environmental changes. A large amount of spatial data is required for managing the disasters and to assess their environmental impacts. Earth observation data offers independent coverage of wide areas for a broad spectrum of crisis situations. It provides information over large areas in near-real-time interval and supplementary at short-time and long-time intervals. Therefore, remote sensing can support disaster management in various applications. In order to demonstrate not only the efficiency but also the limitations of remote sensing technologies for disaster management, a number of case studies are presented, including applications for flooding in Germany 2013, earthquake in Nepal 2015, forest fires in Russia 2015, and searching for the Malaysian aircraft 2014. The discussed aspects comprise data access, information extraction and analysis, management of data and its integration with other data sources, product design, and organisational aspects.

Keywords: Satellite Based Crisis Information, Environmental Changes, Natural Disaster, Man-made Disaster

1. Introduction

The impact of disasters on the environment has become more severe over the last decades. Moreover, the reported number of disasters has dramatically increased, as well as the costs to the global economy and the number of people affected (see Figure 1 for natural disasters) [1,2]. The reasons for these disasters are manifold, and the impact can be found in the increasing vulnerability of societies, infrastructure, and population. Furthermore, extreme weather events have become more common and severe [3].

The increasing occurrences of natural and man-made disasters lead to a growing demand for up-to-date geographic information, especially timely material on rapidly evolving events. This includes comprehensive, near-real-time Earth observation data, which offer independent coverage of wide areas for a broad spectrum of civilian crisis situations [4]. Satellite imagery can serve as a source of information in disaster situation. Accordingly, remote sensing can provide information on various domains of the disaster management, from risk modelling and vulnerability analysis to early warning and damage assessment [5].

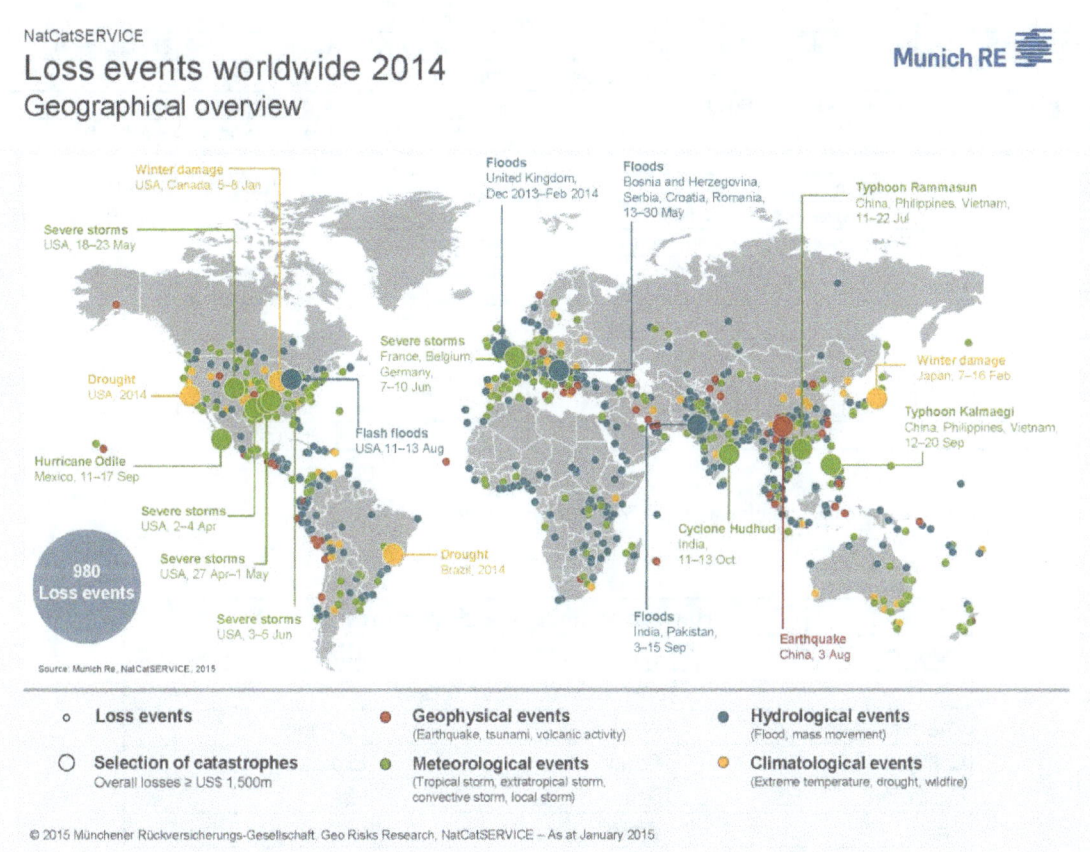

Figure 1. World Map of Natural Disasters 2014 [2]

2. Disaster management and remote sensing

2.1. Disaster types and their environmental impact: A brief overview

There are several ways to classify disaster types [1,6]. One common classification is natural and man-made disasters. Severe geo-physical or climatic events, such as volcanic eruptions, floods, cyclones and fires that threaten people or property, are termed as natural disasters. Man-made disasters are events which are caused by human activities (e.g. industrial chemical accidents and oil spills). Sometimes, natural disasters that are accelerated by human influence are termed human-induced disasters [6]. In addition, the Centre for Research on the Epidemiology of Disasters [7] divides the natural disaster category into six sub-groups, which in turn include 17 disaster types, and 33 sub-types (see Table 1). The technological disaster category is segregated into three sub-groups which in turn include 15 disaster types (see Table 2). Besides, disasters can be categorised as acute (e.g. earthquake) or slow (e.g. drought) based on their onset.

Natural disaster sub-group					
Climatological	Geophysical	Hydrological	Meterological	Biological*	Extraterrestrial*
Natural disaster types and sub-types					
Drought	Earthquake	Flood	Storm	Animal accident	Impact
Glacial Lake Outburst	Ground Shaking	Coastal food	Extra-tropical cyclone	Insect infestation	Airburst
Wildfire	Tsunami	Riverine flood	Tropical cyclone	Grasshoper	Space Weather
Forest fire	Mass movement	Flash flood	Convective Storm	Locust	Energetic particles
Land fire	Volcanic activity	Ice jam flood	Extreme temperature	Epidemic	Geomagnetic storm
	Ash fall	Landslide	Cold wave	Viral disease	Shockwave
	Lahar	Avalanche (snow, debris, mudflow, rockfall)	Heat wave	Bacterial disease	
	Pyroclastic flow	Wave action	Severe winter conditions	Parasitic disease	
	Lawa flow	Rogue wave	Fog	Fungal disease	
		Seiche		Prion disease	

Table 1. Natural disasters categorisation after [7]; *not being considered below

Man-made disaster sub-group		
Industrial accident	Transport accident	Miscellaneous accident
Man-made disaster types		
Chemical spill	Air	Collapse
Collapse	Road	Explosion
Explosion	Rail	Fire
Fire	Water	Other
Gas leak		
Poisoning		
Radiation		
Other		

Table 2. Man-made disasters categorisation after [7]

There are many effects that result from disasters, whether natural or man-made. For instance, the impacts of disasters have a human and an environmental dimension. UNEP concludes that 'environmental conditions may exasperate the impact of a disaster, and vice versa, disasters tend to have an impact on the environment' [8, p.1]. Reference [9] discusses down the environmental impacts for different types of disasters in detail. They raise the interesting point that while most environmental impacts are negative, some are positive. For example, 'floods can help rejuvenate floodplain vegetation and are important drivers of many ecological processes in floodplains' [9, p. 55]. In Table 3, a selection of environmental impacts for different types of disasters is listed.

Natural disasters	Climatological	Drought	Droughts generally damage ecological systems: depletion of water resources, loss of plant and animal life, deterioration of soil, fire
		Glacial lake outburst	Flooding, destruction of plant life, landslide, erosion
		Wildfire	Loss of plant and animal life, erosion, flooding, mud slides, long-term smog
	Geophysical	Earthquake	The dominant losses from earthquakes and mass movements are to structures and potentially to humans. Nevertheless, both disasters can also result in adverse environmental consequences: flora and fauna damaged by the shocks, shifts in land surfaces, alterations in local hydrologic systems.
		Mass movement	
		Volcanic activity	Loss of plant and animal life, deterioration of soil, air and water pollution, long-term smog
	Hydrological	Flood	Major floods have varied effects on river-floodplain ecosystems: e.g. negative impact on trees if they are too long submerged; polluted water infiltrate floodplains and contaminate ground water aquifers; positiv impact like rejuvenate floodplain vegetation
		Landslide	Destruction/loss of plants, erosion, depletion of water resources
		Wave action	Modifies the dynamics of coastal marine communities e.g. the influence the structure of biological communities on rocky shores
	Meterological	Storm	The dominant losses from storms are to structures and potentially to humans. The environmental effects are: e.g. destruction of plants, forest fire, flash flood.
		Extreme temperature	Avalanche, snow melt, loss of plants and animals, flash floods, flooding, drougth, erosion, fire
		Fog	Loss of plant and animal life, decreasing the UV radiation, damages to health for humans, animals and vegetation
Man-made disasters	Technical	Chemical spill Collapse Explosion Fire Gas leak Poisoning	There are several effects on the environment depending on the detailed subtype and dimension of the disaster. Some general impacts are for example: Negative health outcomes from accidental releases of toxins, Loss of plants and animals, Water/soil/air pollution, Damages to health for humans, animals and vegetation,
	Transport	Air Road Rail Water	Destruction of plants, Erosion, Flooding, Fog,
	Miscelleanous	Collapse Explosion Fire	etc.

Table 3. Disasters and their environmental impacts [modified after 9]

The impact of disasters can be reduced through a proper disaster management [10]. The process of disaster management is often interpreted as a cycle consisting of four main phases: mitigation, preparedness, response and recovery (see Figure 2 and for more information [10]).

2.2. Earth observation for disaster management: potential and limitations

The Earth is being imaged each day by a constellation of remote sensing satellites. Two complementary types of Earth observation satellites are particularly relevant to disaster management. 'Geostationary Earth observation satellites' are placed at an altitude of approximately 35,800 kilometres. At this altitude, one orbit takes 24 hours, the same length of time as

Figure 2. Disaster management cycle

the Earth requires to rotate once on its axis [11,12]. In effect, it means that satellites in this orbit remain stationary above the ground and view the whole Earth disk below. Their spatial data resolution is very low but is collected at the same point every 15 minutes. With these kinds of data, the evolution of atmospheric phenomena can be observed, ensuring real-time coverage of meteorological events such as severe local storms and tropical cyclones [13]. The importance of this capability has been exemplified during several hurricane events.

The great advantage of 'polar-orbiting satellites' is the provision of relatively high spatial resolution data (up to 0.3 meter for optical imagery and 1 meter for radar imagery) [11], which is very important for mapping disaster damages in detail, such as affected infrastructure or buildings after an earthquake [13]. Most of the Earth observation satellites are in a low and 'near-polar' orbit with an orbital period of approximately 90–100 minutes and an orbit inclination near 90 degrees. This allows the satellite to see virtually every part of the Earth as the Earth rotates underneath it. However, no spot on the Earth's surface can be sensed continuously or at any point of time from a satellite in a polar orbit. The time elapsed between observations of the same point on the Earth (revisit time) is limited to once every few days with the same sensor parameters or maximum once a day for steerable satellite. Moreover, most satellites do not continuously collect data due to limitations in power and memory. Some offer regular and reliable data acquisition while others may be more ad hoc, collecting only 5 or 10 minutes' worth of data in a 90-minute orbit. Data are stored on board the satellite until it is in sight of a ground station to downlink the data. The time between an image being taken and being available to download can range between a month to a few minutes and is getting faster all the time. Thus, the collection of high-resolution data has some limitations regarding acquisition time, data provision and image extent.

The Earth observation satellites have their own special systems of imaging sensors which make use of the visible, infrared, microwave and other parts of the electromagnetic spectrum [11]. The characteristics of some sensors that are commonly used to support disaster management are listed in Table 4.

Data type	Sensor	Nadir spatial resolution (m)	Bands	Swath (km)	Revisit Frequency
Optical (multispectral)	Worldview-3	0.31 1.24 3.7 30	Panchromatic 8 Multispectral 8 SWIR 12 CAVIS (Corrects for Clouds, Aerosols, Vapors, Ice & Snow)	13.1	1.1 days at 1 m GSD or less 4.5 days at 20° off-nadir or less
	Worldview-2	0.46 1.84	Panchromatic 8 Multispectral	16.4	1.1 days at 1 m GSD or less 3.7 days at 20° off-nadir or less
	Pleiades-1A / 1-B	0.70 2.00	Panchromatic 4 Multispectral	20	Daily
	SPOT-6 / -7	1.50 6.00	Panchromatic 4 Multispectral	60	Daily
	RapidEye	6.5	5 Multispectral	77	Daily
Thermal	ASTER	15 30 90	4 Multispectral 6 SWIR 5 TIR	60	4-16 days
	MODIS	250 500 1,000	36 bands (VIS, NIR, SWIR/MWIR, LWIR)	2,330	Daily
Synthetic Aperture Radar (SAR)	TerraSAR-X / TanDEM-X	1 3 18	Spotlight Stripmap ScanSAR	10 30 100	11 days
	Cosmo-SkyMed	<1 3-15 30-100	Spotlight Stripmap ScanSAR	10 40 100-200	1.5 days
	Radarsat -2	3 25 8 8 25 25 50 100	Ultra-fine Fine Quad-pol fine Standard Quad-pol standard ScanSAR narrow ScanSAR wide Extended high	20	Every few days
	ALOS	10 100	PALSAR (Fine) PALSAR (ScanSAR)	40-70 250-350	Several times per year as per JAXA acquisition plan

Table 4. Examples of sensors and their characteristics to support disaster management [modified after 16]

'Optical data' are of great importance for disaster management support, because they can be used nearly for all disaster types and for all phases of disaster management. For example, they are used for planning the logistics of relief actions in the field immediately after an earthquake or tsunami [13–14]. Optical images are easy to understand and interpret even for non-specialists, particularly when it consists of the three visual primary colour bands (red, green and blue) and the bands are combined to produce a 'true colour' image. However, the interpretation of false colour composite images is not intuitive and requires expert knowledge; likewise, all advanced analysis techniques need comprehensive know-how. To select the most appropriate data type for the needs of the individual disaster situation, the characteristics of the sensor are of great importance [15]. Particularly, temporal and spatial resolutions are key factors. For example, for mapping an earthquake in an urban area optical data with a spatial resolution of <0.5 meters are most valuable. The most crucial point for the use of optical images is their availability. Due to cloud coverage, haze and other atmospheric conditions useful optical images could not obtained by every satellite overpass. Aggravating this situation, there are some disasters such as wildfire or severe storms which are characterised by clouds and smoke.

The 'thermal imagery' offers excellent possibilities for automated extraction of anomalous high temperature or hot spots caused by wild fires or information about volcanic eruptions. However, due to the fact that energy decreases with increasing wavelength, thermal wavelength have relatively low energy levels and consequently thermal image data have a lower spatial resolution than optical data. [16]. Techniques for automatic fire detection from the space are operational and are accepted by the users (e.g. European Forest Fire Information System) [17].

'Microwave sensors' are of great value for the fast response mapping and analysis tasks, as they allow imaging at wavelengths almost unaffected by atmospheric disturbances such as rain or cloud. Most modern synthetic aperture radar (SAR) sensors are designed to acquire data from various ground resolution elements (see Table 4). In most applications, only the relative variability of backscatter intensity within the image is used. Nonetheless, backscatter intensity and the phase of SAR images can be utilised. Phase information of a single SAR data set has no value, but the comparison of phases between two SAR images acquired at distinct times are utilised in SAR interferometry or INSAR. Moreover, with modern satellites (e.g. TerraSAR-X and Radarsat-2) it is possible to acquire simultaneous data with more than one polarisation [16]. SAR systems can be used to map flooding or to measure earth deformations before and during earthquakes or volcanic eruptions, particularly when post-event imagery can be jointly analysed with archived reference imagery for change detection or interferometric coherence or displacement measurements [15, 13].

Table 5 summarises the remotely sensed data types and image processing techniques for information extraction about natural disasters.

In general, the availability of appropriate data with respect to acquisition time, image extent, spatial as well as temporal and spectral resolution is an important consideration for most applications in the disaster context [18]. Particularly, there are numerous examples for the importance of the necessity of fast availability of remote sensing data like damage assessment maps for earthquakes, landslides or flooding. However, for monitoring the spread of an oil spill or the extent of flooding the revisit time is relevant too [13].

Remote sensing has proven to be useful for a range of applications. Especially high spatial resolution data and remote sensing techniques are being deployed in the context of the disaster

Data type	Sensor examples	Technique	Application
Optical (multispectral)	Worldview, Pleiades, Quickbird, Ikonos, RapidEye, SPOT, ASTER, Landsat, ALOS	Manual interpretation	Infrastructure and property damage due to flooding, earthquakes, landslides, etc.
		Spectral classification	Location and extent of flooding, landslides, volcanic debris, fire scars
		Semivariogram analysis and other textural classifiers	Damage due to earthquakes; location of landslides
		Image thresholding (including band ratios)	Location and extent of flooding, landslides, volcanic debris, fire scars
		Image differencing	Location and extent of flooding, landslides, volcanic debris, fire scars
		Postclassification change detection	Location and extent of flooding, landslides, volcanic debris, fire scars
		DEM generation	DEM is used as a supplementary information in variety of studies
Thermal	ASTER, MODIS, AVHRR	Split window	Crater lake temperatures, lava flow, precursor to earthquake activity, temperature and size of fire hotspots
		Dual band	Crater lake temperatures, lava flow, precursor to earthquake activity, temperature and size of fire hotspots
Synthetic Aperture Radar (SAR)	TerraSAR-X, TANDEM-X, Cosmo-SkyMed, Radarsat-1/2, JERS-1, ERS-1/2, ENVISAT, ALOS	Coherence	Coherence Change detection due to landslide, flooding, fire, etc.
		Backscatter intensity	Coherence Change detection due to landslide, flooding, fire, etc.
		Interferometry/DEM generation	Change detection due to landslide, flooding, fire, etc.
		Differential interferometry	Surface deformation due to volcanic or tectonic activity; velocity and extent of slow moving landslides
		Polarimetry	Landcover classification and change detection

Table 5. Remotely sensed data types and image processing techniques for information extraction about natural disasters [modified after 13]

management domain, from risk modelling and vulnerability analysis to early warning and damage assessment (see Table 6). A broad assessment of several remote sensing sensors (optical, thermal, SAR, etc.) and their utility for providing information about natural disasters is given in Ref. [13, p. 200–201].

Reference [19, p. 2-3] concludes that 'the most evident parts are preparedness (warning for storms, cyclones, floods, etc.,) and response (mapping of all types of crisis impact and situations), while applications of satellite information during the phases of recovery and mitigation prevention are being still further developed'. Additionally, the authors give the following main reasons for a drastically increased demand for rapid satellite data analysis for all kinds of disaster and phases over the past years:

- accessibility of very high resolution optical (up to 0.3 meter) as well as radar imagery (up to l meter) from space has risen significantly over the past years even for the civilian domain

- relief agencies rapidly gain a better understanding on what these new geoinformation technologies can bring to their work in the fields of mission planning, logistics, situation awareness and even mission security

			Mitigation	Preparedness	Response	Recovery
Natural disasters	Climatological	Drought	Risk modelling; vulnerability analysis; land and water management planning	Weather forecasting; vegetation monitoring; crop water requirement mapping; early warning	Monitoring vegetation; damage assessment	Informing drought mitigation
		Glacial Lake Outburst	Mapping glacial lake outburst-prone areas; glacial monitoring; delineating flood-plains; land-use mapping	Weather forecasting; glacial monitoring, lake outburst detection; early warning	Flood mapping; evacuation planning; damage assessment	Damage assessment; spatial planning
		Wildfire	Mapping fire-prone areas; risk modelling	Fire detection; predicting spread/direction of fire; early warning	Coordinating fire fighting efforts	Damage assessment
	Geophysical	Earthquake	Building stock assessment; hazard mapping	Measuring strain accumulation	Planning routes for search and rescue; damage assessment; evacuation planning; deformation mapping	Damage assessment; identifying sites for rehabilitation
		Mass movement	Hazard mapping	Measuring strain accumulation	Planning routes for search and rescue; damage assessment; evacuation planning; deformation mapping	Damage assessment; identifying sites for rehabilitation
		Volcanic activity	Risk modelling; hazard mapping; digital elevation models	Emissions monitoring; thermal alerts	Mapping lava flows; evacuation planning	Damage assessment; spatial planning
	Hydrological	Flood	Mapping flood-prone areas; monitoring fuel load; delineating flood-plains; land-use mapping	Flood detection; early warning; rainfall mapping	Flood mapping; evacuation planning; damage assessment	Damage assessment; spatial planning
		Landslide	Risk modelling; hazard mapping; digital elevation models	Monitoring rainfall and slope stability	Mapping affected areas; and slope stability	Damage assessment; spatial planning; suggesting management practices
		Wave action	Risk modelling; vulnerability analysis	Wave action detection; early warning	Mapping affected areas	Damage assessment; spatial planning
	Meterological	Storm	Risk modelling; vulnerability analysis	Early warning; long-range climate modelling	Identifying escape routes; storm surge predictions; cyclone monitoring; impact assessment; crisis mapping	Damage assessment; spatial planning
		Extreme temperature	Risk modelling; vulnerability analysis	Weather forecasting; long-range climate modelling; early warning	Crisis mapping; evacuation planning; damage assessment	Damage assessment; spatial planning
		Fog	Risk modelling; vulnerability analysis	Weather forecasting; long-range climate modelling; early warning	Crisis mapping; evacuation planning; damage assessment	Damage assessment; spatial planning

Table 6. Remote sensing applications for disaster management [complemented after 12]

- media and the public raise the demand for up-to-date easy-to-understand visual information on disaster areas and ongoing relief work

The following sections focus primarily on the contribution of remote sensing to the response phase, in particular, giving a brief overview of the workflow from an emergency call or request for assistance, through satellite tasking, data acquisition, analysis, map provision and furthermore explaining some existing operational services, and finally a number of case studies are presented.

2.3. Rapid mapping workflow

No decision maker or relief worker can work with raw satellite imagery. To generate the required situation maps, reports or statistics, which can be read and understood by non-satellite expert users, experts in remote sensing and cartography are necessary. In 2004, German Aerospace Center (DLR) was one of the first institutions, which has set up a dedicated interface called Center for Satellite Based Crisis Information (ZKI) to facilitate the use of its Earth observation capacities in the services of national and international response to disaster situations [18]. ZKI's function is particularly 'rapid mapping' – the rapid acquisition, processing and analysis of satellite data and the provision of satellite-based information products. Analyses are tailored to meet the specific requirements of national and international political bodies or humanitarian relief organisations. In order to provide up-to-date and relevant satellite-based cartographic information and situation analysis, it is necessary to establish efficient and operational data flow lines between satellite operators, receiving stations and distribution networks on the one hand and the decision makers and relief workers on the other hand. Service lines and feedback loops have been created to allow best possible data and information provision, as well as optimised decision support [20]. In order to meet with users' demands and service requirements in crisis situations, ZKI set up a rapid mapping workflow (Figure 3) ensuring a fast access to available, reliable and affordable crisis information worldwide.

Schedules for the full cycle from the emergency call (mobilisation phase), satellite tasking (data acquisition), pre-processing, analysis and interpretation, map production and data provision to the end-user are tight (as fast as possible). Hence, rapid mapping is still a complex task [15].

After the mandatory decision process, whether satellite analysis is appropriate for the respective crisis or not, the area of interest has to be defined and cross-checked to avoid false geolocation. Following this iterative process, it has to be assured that all applicable satellites are programmed for data acquisition. Furthermore, an enquiry for corresponding archive imagery has to be set up for documentation of the pre-disaster situation and change detection analysis. Besides the procurement of satellite data, it is necessary to check and prepare supplementary geodata such as population and infrastructure data, road network, contour lines and administrative boundaries. Experience of several activations and user feedback shows that additional geoinformation increases the satellite data analysis significantly. This includes place names, critical infrastructure, transportation network or further detailed specifications. Availability and access to accurate and up-to-date spatial data, particularly in remote regions, are the most crucial problems [18].

After receiving the archived and recently recorded satellite imagery, essential pre-processing has to be done. This includes geo- and ortho-rectification as well as radiometric corrections and data format conversions. Data re-projection is necessary due to varying demands and standards. In the majority of activations, a Universal Transverse Mercator (UTM) projection is used due to global applicability and following international standards. Depending on user's needs, crisis type and extent, different analysis process chains have to be applied [18].

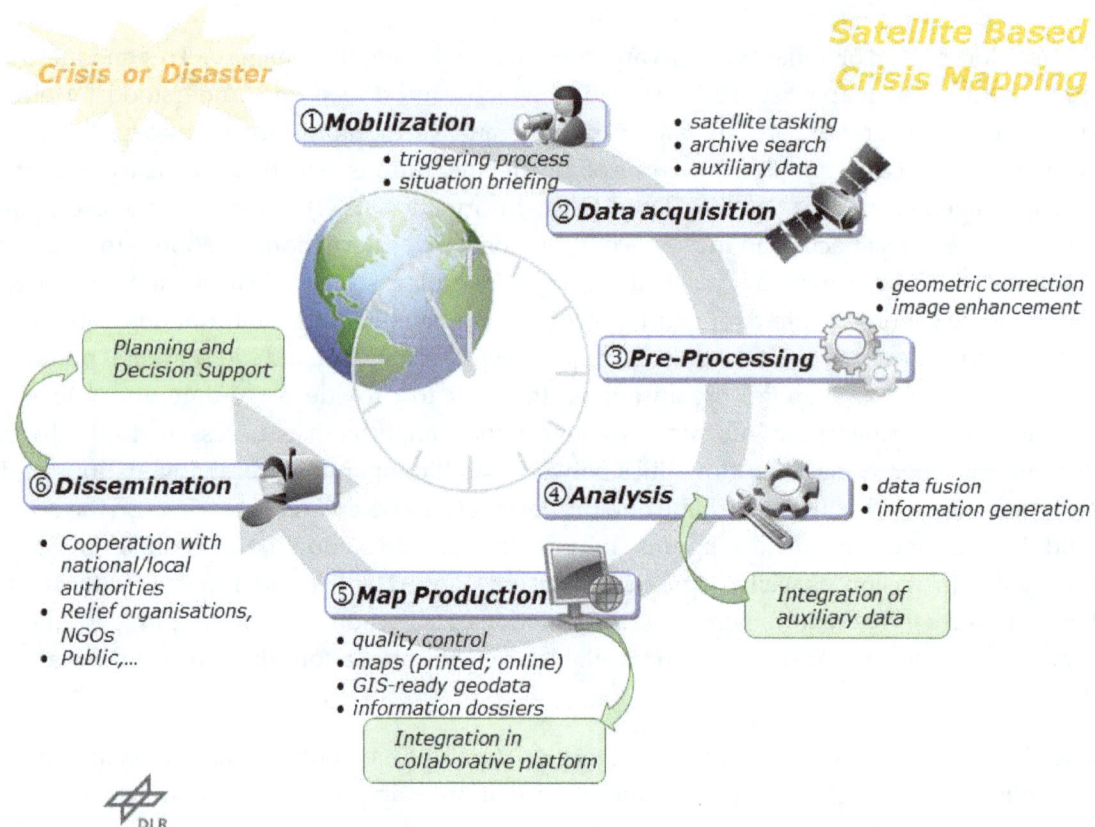

Figure 3. Rapid mapping workflow

Derivation of water surfaces or general damage assessment is dependent on input data type, scale and possible availability of archived satellite imagery. Before and after image comparison allows the quantification of affected areas. This change detection method can either be applied for optical or radar imagery in order to detect areas where significant change can be identified. Furthermore, general image classification and differencing methods allows quantification of flooded areas, fire scars or damaged areas [19].

Situation and damage maps are generated in order to translate complex satellite information in readable and coherent crisis information. Following this map compilation, an adapted map generation process is applied. A settled quality control process takes place after each single product generation step as well as before publishing. Delivery is accomplished via Internet, intranet, ftp, e-mail or satellite communication. Furthermore, printed and laminated maps will be sent via express delivery on request. User feedback from field units has proved to be an important source for optimisation. Maps are updated when new and improved data are available or knowledgeable feedback is received even though the maps are published and delivered [18].

In order to fulfil its tasks, DLR-ZKI is involved in international, European and national mechanisms providing space-based information supporting the disaster relief (e.g. International Charter Space and Major Disaster). The understanding of the organisational frameworks of these mechanisms, their activation procedures and workflows are a prerequisite to take advantage of the products provided by these mechanisms.

2.4. Mechanisms of providing satellite-based crisis information

2.4.1. International Charter 'Space and Major Disasters'

For providing fast and reliable image access on archive or new post-event imagery effectively, there is a need for more than a single research-oriented or commercial system. Thus, 'effective and well-balanced coordination among the different observing systems is required in order to allow best service to the civil-protection and humanitarian relief community' [18, p. 1527].

With the installation of International Charter Space and Major Disasters in 1999, from now onwards referred to as 'Charter', a globally functioning mechanism was established to provide a unified system of rapid space data acquisition and delivery in case of natural or man-made disasters [21]. The Charter is a consortium of space agencies and satellite data providers. Each member agency of the Charter has committed resources to support relief organisations as well as civil protection and defence organisations with free of charge satellite (raw) data in order to help mitigating effects of disasters on man life and environment. Its members, conscious of the need to improve its access globally, have adopted the principle of 'Universal Access': any national disaster management authority will be able to submit requests to the Charter for emergency responses [21]. Proper procedures have to be followed, but the affected country does not have to be its member as it was before. A registration process is available for national authorities to express interest in participating in the Charter. Universal Access implementation started in September 2012 and is being implemented gradually [21].

Since its inception in 2000, the Charter has been activated for more than 470 disasters (as end of September 2015), in more than 110 countries. In 2014, the Charter was activated 41 times for disasters in 30 countries. In the same year, more than 75% of Charter activations were based on weather-related disasters such as flooding, ocean storms and landslides, while solid Earth-related hazards (e.g. earthquakes, volcanic eruptions) represented 10% of Charter activations; activations for man-made disasters (e.g. oil spills) are marginal (<5%; see Figure 4) [21].

Comparing Charter activations with occurrence of disasters of hazard types reported by emergency events database (EM-DAT), proportions of both fit together to some extent (see Figure 5). One obvious difference can be recognised in category "Others" which incorporates particularly all man-made disasters. Nevertheless, the Charter covered 7 of the 10 most severe disasters by fatalities 2014 as reported by EM-DAT (Table 7).

Top 10 Disasters – Number Killed – 2014					
Country	Disaster type	Date	#killed	#Affected people	Total Damage (000' $)
China P Rep	*Earthquake*	*3/8/2014*	*731*	*1,120,513*	*5,000,000*
Nepal	*Landslide*	*2/8/2014*	*450*	*184,894*	-
Afghanistan	*Flood*	*24/4/2014*	*431*	*140,100*	-
Pakistan	*Flood*	*1/9/2014*	*367*	*2,470,673*	*2,000,000*
India	*Flood*	*8-9/2014*	*298*	*275,000*	*16,000,000*
Nepal	Flood	12/8/2014	217	28,279	-
India	Landslide	30/7/2014	209	-	-
Sri Lanka	*Landslide*	*29/10/2014*	*196*	-	-
China P Rep	*Storm*	*7/4/2014*	*128*	-	-
Philippines	Storm	15/7/2014	111	4,654,966	820576

Table 7. Ten most severe natural disasters by number of fatalities in 2014 based on EM-DAT statistics [7] and events covered by Charter activations (indicated in bold and italics) [source 21; p.49]

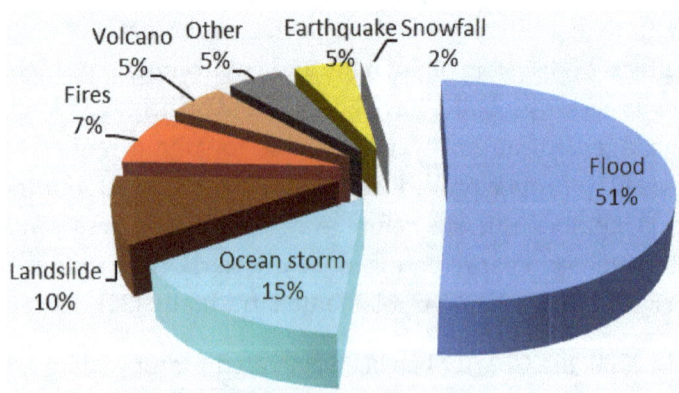

Figure 4. Percentage of hazard-type Charter activations in 2014 [21, p.24]

Due to user feedbacks and meaningful statistics, it can be concluded that a meaningful satellite observation information capacity was established for a variety of non-expert users.

However, it should be mentioned, that the Charter does not concern the whole disaster management cycle (see Figure 2) and not for long humanitarian crisis as well. Moreover, the rapid mapping value-adding activities (see the following section) are not primary in the mandate of the Charter. The analysis of the satellite (raw) data and map production are often entrusted to associated value-adders. Today there is no other operational capacity playing such

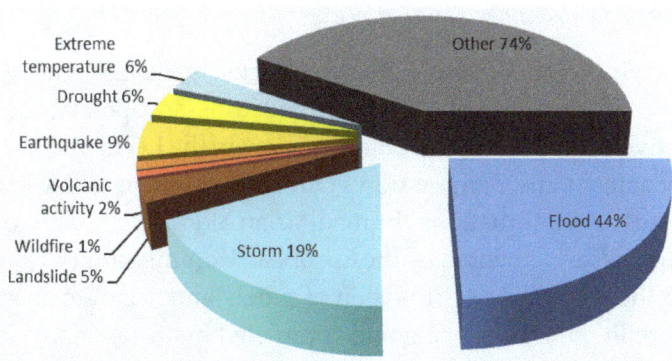

Figure 5. Occurrence of disasters in 2014 in percentage, by hazard type [data source 7]

an important role on a worldwide basis. However, other space-based initiatives are available providing new capacities to other users.

2.4.2. Sentinel Asia

Another collaboration between space agencies is the so-called Sentinel Asia initiative. It has a regional focus and was established in 2005, as a collaboration between regional space agencies and disaster management agencies, applying remote sensing and Web-GIS technologies to assist in disaster management in the Asia-Pacific region. Until today multiple national agencies of about 25 countries in the region have joined and benefited from the disaster support services provided by Sentinel Asia. It intends to expand efforts like the Charter and make relevant data available to all countries and many more people in the region [22].

Sentinel Asia also cooperates with the Charter: since its inception Sentinel Asia provides a regional enhancement to the Charter, as it allows any country in the region to join their network and request disaster-relevant information, regardless of their membership of the Charter (even before the Universal Access was implemented). Moreover, Sentinel Asia built up an expert team with different knowledge base, such as disaster management agencies, space agencies, as well as relevant regional and international entities. They created a network with so-called data provider nodes (DPNs), where several regional space agencies and related institutes providing satellite data from their national satellite systems to the so-called data analysis nodes (DANs) [23]. These DANs analyse raw image data together with their own geospatial data. Moreover, they implemented specific technical working groups, which aim to accelerate and optimise information analysis process (e.g. expand utilisation of satellite-derived products for tsunamis or wildfires). In parallel with the activities above, capacity building for technical and emergency-response agencies users of the Sentinel Asia system is realised [22–23].

In summary, Sentinal Asia is a direct and active collaboration with regional disaster management agencies, and great regional network of data providers, data analysis nodes and users. With regard to the fact that Asia comprises 39% of the worldwide total disasters, Sentinel Asia is a very valuable initiative [22].

2.4.3. Copernicus Emergency Management Service

Yet another service which collaborates with the Charter is the Copernicus Emergency Management Service (EMS). Copernicus EMS is intended as an operational service offered to authorised users active in the field of crisis management in the EU member states, the European civil protection mechanism, the Commission's Directorates-General (DGs) and the participating executive agencies and international humanitarian aid [24]. The service started its operations in April 2012 and is implemented by the European Commission DG Joint Research Centre (JRC). The scope of the service is the provision of timely and accurate geospatial information derived from the satellite remote sensing and completed by the available in situ or open data sources. Copernicus EMS is provided free of charge for the users, during all phases of the disaster management cycle, and in two temporal modes [25].

EMS service and all other Copernicus services such as land monitoring or atmosphere monitoring are based on the provision of satellite imagery from contributing missions that are made available through the Copernicus Space Component Data Access (CSCDA) system operated by European Space Agency (ESA) since 2008. In future, the service will also be supported by all the Sentinels; for Sentinel 1-A first maps were already produced [26].

Analysis products are standardised and depend on the set of parameters chosen by users when placing the service request. For rapid mapping, the following product categories are offered: reference maps, delineation maps (providing an assessment of the geographic extent of the event) and grading maps (providing an assessment of the damage grade and its spatial distribution) [25].

Unlike the Charter analysis and map, production is explicitly within the mandate of Copernicus EMS. Therefore, an agreement has been set up to exploit the advanced crisis mapping capability of the EMS to support Charter requests pertinent to European policy sectors. Another important advantage of Copernicus EMS is the opportunity to request geospatial information in support of disaster management activities not related to immediate response. This is of particular importance for activities dealing with prevention, preparedness, disaster risk reduction and recovery phases [25]. For this purpose, there are three categories of maps offered: reference maps, pre-disaster situation maps and post-disaster situation maps.

In summary, Copernicus EMS is a fully operational service with a predefined and standardised product portfolio, covering the whole disaster cycle, which is free of charge for authorised users. In contrast, satellite data providers as well as value adders have paid service contracts with the European Commission, which leads often to a faster and guaranteed product delivery, but not necessarily to better products. Some restrictions are given for the service: only large-scale disasters and crises are within the scope of the service and the request should not be related to an existing on-going conflict or crisis with EU military operations or in politically sensitive areas [25].

2.4.4. UNITAR Operational Satellite Applications Programme (UNOSAT)

UNOSAT is the United Nations Institute for Training and Research (UNITAR) Operational Satellite Applications Programme which was created in 2000 [27]. UNOSAT provides maps, reports as well as geographic information system (GIS) compatible data layers for natural hazards, complex emergency situations or conflict crises – at no cost to the user. The users are entities of the United Nations systems such as OCHA, UNHCR, UNICEF, WFP, UNDP, WHO, IFRC, ICRC; International and national NGOs and the governments of affected countries. UNOSAT is covering the response and recovery phase and is working worldwide. UNOSAT collaborates with several partners (e.g. other services, satellite data providers, UN entities, companies like Google and ESRI) [28].

2.4.5. SERVIR

SERVIR mechanism is a joint venture between NASA and the United States Agency for International Development (USAID) [29]. It integrates satellite observations, ground-based data and forecast models to help developing nations in Central America, East Africa and the Himalaya region to assess environmental threats and to respond to and assess damage from disasters of natural origin. SERVIR is a multi-agency and multi-government mechanism with over 30 partners and collaborators and is endorsed by governments in Central America, Africa and the Hindu-Kush Himalaya region of Asia. The coordination office is located in United States and is supported by three regional centres: The Regional Centre for Mapping of Resources for Development (RCMRD) in Kenya, the International Centre for Integrated Mountain Development (ICIMOD) in Nepal and the Water Centre for the Humid Tropics of Latin America and the Caribbean in Panama [28].

Program supports not only national governments, but also universities, non-governmental organisations, and the private sectors. Users of SERVIR are government officials, disaster managers, scientists/researcher, students and the general public [29]. SERVIR serves as a source for satellite imagery and information provider during extreme events. The SERVIR mechanism is intended to respond to needs for satellite-based geoinformation in Mesoamerica, Africa or the Himalaya [28].

2.4.6. ZKI service for federal agencies

One of the first national operational services providing rapid space data acquisition and delivery in case of natural or man-made disasters is the so-called ZKI Service for Federal Agencies (ZKI-DE). ZKI-DE was established in January 2013, based on a framework contract between German Federal Ministry of the Interior (BMI) and the German Aerospace Center (DLR), coping all phases of the disaster management cycle. It enables German national authorities and other authorised users to order products of DLRs Center for Satellite Based Crisis Information (ZKI), even for requests at regional scale and for users like national security authorities (with the option of arrangements of confidentiality). Moreover, aiming at a better and more customised use of the products by public authorities, the service includes not only the provision of maps and dossiers in case of a disaster but also user trainings, a consulting service and continuous further developments based on user requirements and new technical capabilities [30]. This cooperation is not limited to BMI and its special agency (e.g. like 'Federal Office of Civil Protection and Disaster Assistance' or 'Federal Criminal Police Office'). As a first institution, the German Red Cross also uses ZKI-DE products for their emergency operations worldwide [30]. As a matter of fact, due to licence and safety regulations products could often not be published.

2.4.7. Other initiatives

There are more and more actors gathering/providing further space-based disaster information such as private companies (e.g. Digital Globe, Google, ESRI). Other interesting and upcoming actors are volunteer organisations like Map Action and crowdsourcing crisis mapper (e.g. Open Street Map initiative–OSM, Tomnod owned by Digital Globe). In this context, data acquisition and reliability are often critical aspects. Depending on the expertise of volunteer's equipment and raw data quality, analysis can vary considerably. Nevertheless, changing data policies in case of (major) disasters and new techniques in image analysis can facilitate the access to satellite data, as well as the dissemination of rapid-mapping products.

3. Examples for applying satellite-based information in disaster management: Case studies

In this section, a number of examples of satellite imagery application for disaster relief intend to highlight swift and synergistic use of state-of-the-art processing techniques and rapid data access. These rapid-mapping results could be achieved by building on existing scientific results, long-term engineering experience in the domain of satellite data processing and last but not least operational data access mechanisms. It is not indented to report major generic methodological research results or method comparison here.

3.1. Flooding in Germany 2013

Extreme flooding in Germany and other parts of the Central Europe began after several days of heavy rain in the late May and early June 2013 [31]. Flooding and damages primarily affected southern and eastern German states as well as Czech Republic and Austria. Switzerland, Slovakia, Belarus, Poland, Hungary and Serbia were affected to a lesser extent [31].

German Joint Information and Situation Centre (GMLZ) tasked Charter, Copernicus GIO EMS (Precursor of Copernicus EMS) and national Service ZKI-DE with the provision of satellite data and the creation of satellite and aerial image-based situation information covering the regions most affected by the current floods in Thuringia, Saxony, Bavaria and Baden-Wuert-temberg [32-34]. GMLZ is a facility of the German Federal Office of Civil Protection and Disaster Assistance (BBK). Charter and Copernicus EMS provided fast and cost-free access to satellite images covering a disaster area; moreover, Copernicus EMS delivered 38 maps (reference and disaster). National ZKI-DE service complemented the response products with up-to-date airborne data and more than 50 products such as supplementing monitoring maps or web services (see Figure 6).

Figure 6. Flooding in Germany 2013 – ZKI-DE Web service [32]

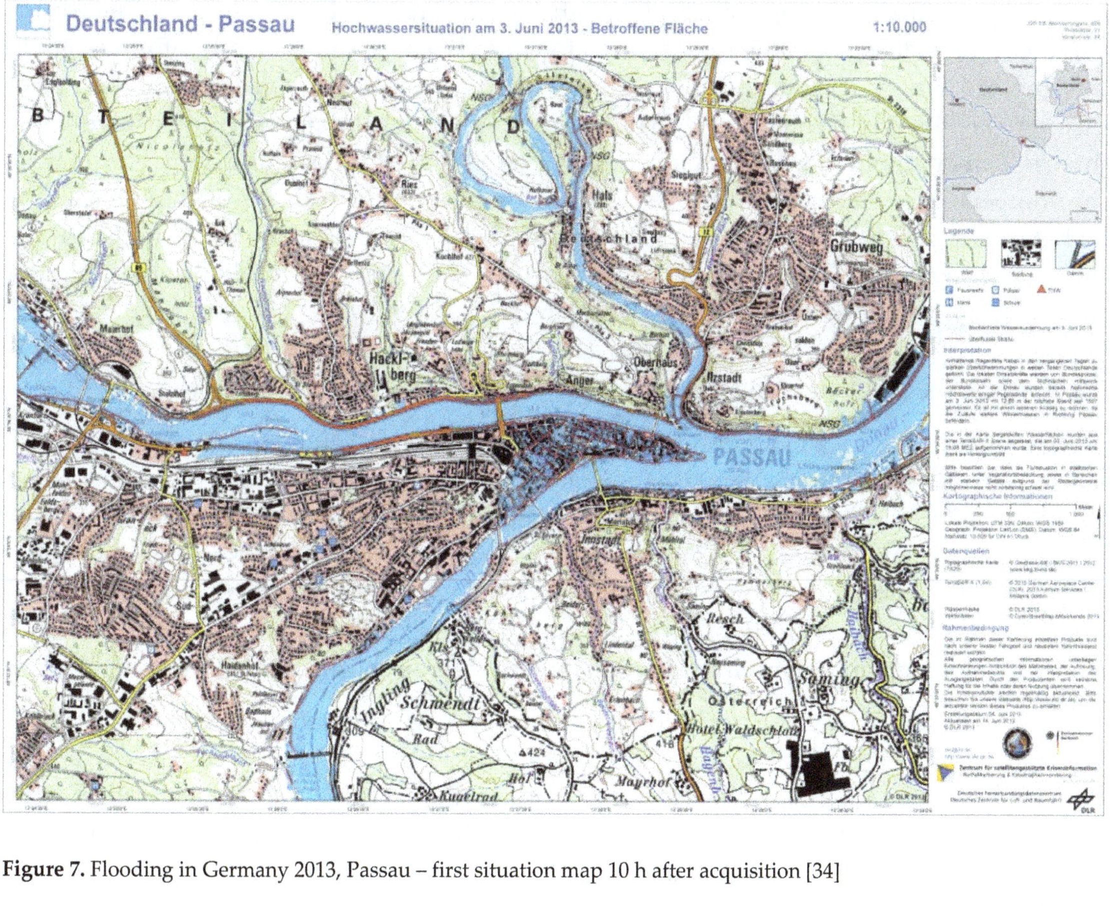

Figure 7. Flooding in Germany 2013, Passau – first situation map 10 h after acquisition [34]

ZKI-DE was able to present the first situation map products to the users. The products based on the German radar-satellite TerraSAR-X and were delivered 10 hours after acquisition. The maps show the flood extent derived from the radar data, as backdrop serves as a topographic map (see Figure 7). Derivation of flood extent from TerraSAR-X data and map production/dissemination just took 4 hours (6 hours for data downlink and pre-processing, respectively). These fast rapid-mapping results could be achieved by building on existing operational organisational structures (24/7) with trained staff, (semi-) automated image analysis procedures, and several templates as well as models/macros for the map products and their dissemination.

Nearly all the products were published for everyone on websites and were used in the disaster response phase by several users in Germany – from decision makers in the situation centres as well as local or top-ranking politician. Moreover weeks and months after the disaster, vector data sets of the disaster extent were requested several times by environmental and research institutes. These entities work in different phases of the disaster management cycle. Following up the experiences of the flooding 2013, several actions were implemented to reduce the impact of such heavy weather conditions [35].

In general, optical as well as radar satellite remote sensing have proven to provide essential large-scale information on flood situations. For optical input data, the standard semi-automatic method is (unsupervised) classified [36]. If the spectral resolution of the sensor and/or the cloud coverage does not allow clear semi-automatic classification, the flood information can be extracted via visual interpretation. Change detection analysis is used if pre- and post-disaster satellite data are available [34]. Even though, optical data provide positive result information on inundation, radar data are a preferred input for flood detection. Fortunately, the number of automatic image processing algorithms to derive flooding from high-resolution SAR data (TerraSAR-X, Radarsat-2, Cosmo–SkyMed) has increased in the last years. One thing in common in these algorithms is that they make use of automatic thresholding algorithms for the initialisation of the classification process [36–39].

3.2. Earthquake and Landslide in Nepal 2015

On 25th April 2015, a 7.8 magnitude earthquake hit the Himalayan region. The epicentre was located near Kathmandu, the capital of Nepal. In addition to Nepal, India, China and Bangladesh were affected. The event and several aftershocks caused wide-ranging destruction. The earthquake triggered several landslides, an avalanche on Mount Everest. More than eight million people were affected by the earthquake [40]. Several actors were involved in producing useful information as a response to the disaster. Copernicus EMS was activated by European Commission's Humanitarian aid and Civil 40 products were published at the EMS webpage [41].

The International Charter on Space and Major Disasters was activated by several organisations (e.g. Indian Space Research Organisation – ISRO), and a plenty of maps were made available on webpage of the Charter in the aftermath of the earthquake [42]. Moreover, Sentinel Asia, ZKI-DE and several other actors [43] delivered a number of standard maps (see Figure 8) and innovative products such as a 3D-Animation flight over Kathmandu before the earthquake, based on 10 cm airborne data (see Figure 9) [42,44]. Even the National Geospatial-Intelligence

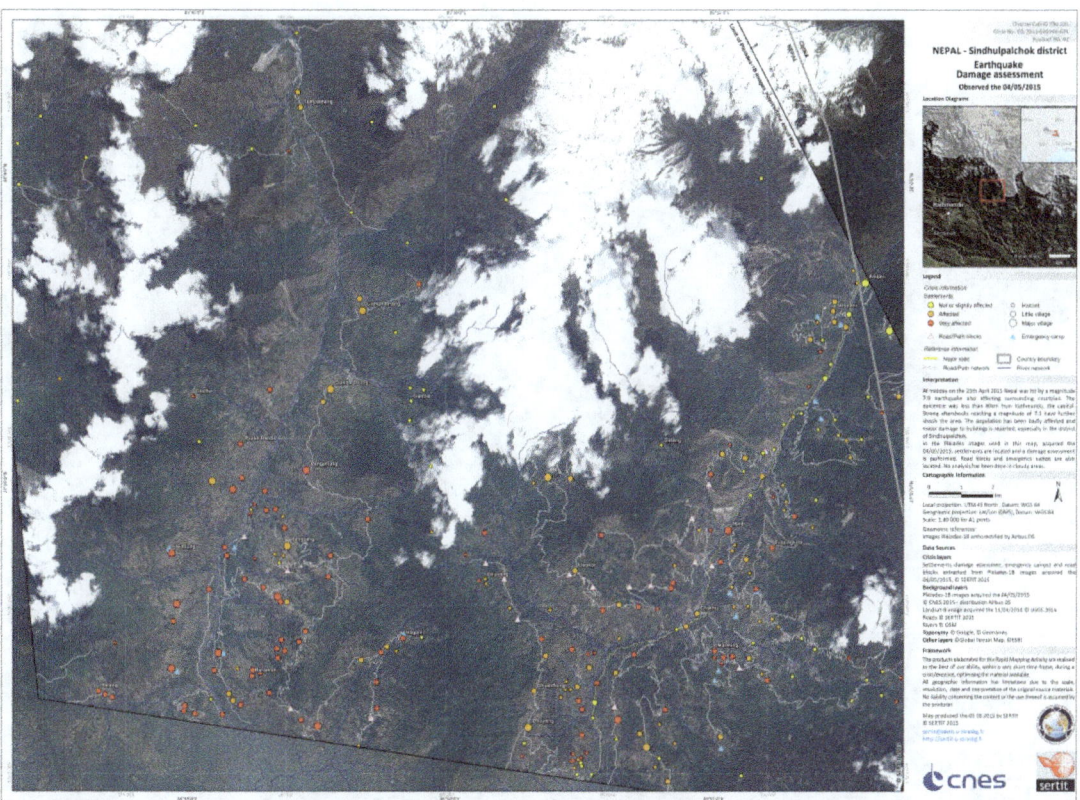

Figure 8. Earthquake in Nepal—Damage assessment map based on Pléiades data and visual interpretation [42]

Agency released unclassified geospatial intelligence data, products and services [42]. Within the crowdsource community also several initiatives were working on the Nepal earthquake such as Tomnod and Open Street Map (see Figure 10), Map Action, Micromappers [45].

Various methodologies have been proposed for earthquake damage assessment using optical and radar Earth observation data. For estimating infrastructural damages based on SAR, methods exploiting changes in backscattering intensity and the related image correlation coefficient [47] or a combination of backscatter intensity, phase changes and/or ancillary data are often used [48]. Reference [15] stated that 'most SAR-based change detection approaches suffer from a lack of archive data with the same acquisition parameters as the post crisis imagery'. Consequently, only very few of the SAR or optical-based approaches have been targeted for the use in an operational rapid mapping environment. For analysing optical satellite data several analysis methods were used to detect damages. Several authors applied either semi- or fully automatic change detection methods for earthquake damage assessment [49,47]. However, for various reasons, such methods have rarely been applied during rapid mapping activities [15]. For instance, automatic change detection approaches will potentially detect changes that are not related to earthquake damages (e.g. vegetation changes, different illumination, etc.). Moreover, in case of cloud coverage, haze or radiometric and spectral problems just manual extraction methods like visual interpretation and grid interpretation can be performed. The

Figure 9. Earthquake in Nepal—Screenshot of a 3D-Animation flight over Kathmandu before the earthquake based on 10 cm airborne data (sensorDLR MACS) [44]

selection of the method depends on the requested information. For instance, if single (small-scale and heterogeneous) objects (e.g. houses, streets, ships and other relevant objects) have to be identified only visual interpretation is possible. Otherwise, the grid analysis allows the interpretation in specified areas (grid cells) and takes the relation of the neighbourhood into account by spatial aggregation of the thematic information content to be provided [18].

In summary, a fast activation of several mechanisms has taken place and resulted in large amounts of satellite imagery and airborne data. Nearly all results are based on optical data and visual interpretation. Many useful maps and visualisations facilitated a general understanding of the situation as well as the assessment of detailed aspects of the disaster and the relief work, including damage overview, road and infrastructure accessibility, gathering areas, strategic holding areas etc. According to several users' feedback, the maps and layers (streets, damage etc.) provided vital information with respect to evacuation planning, general

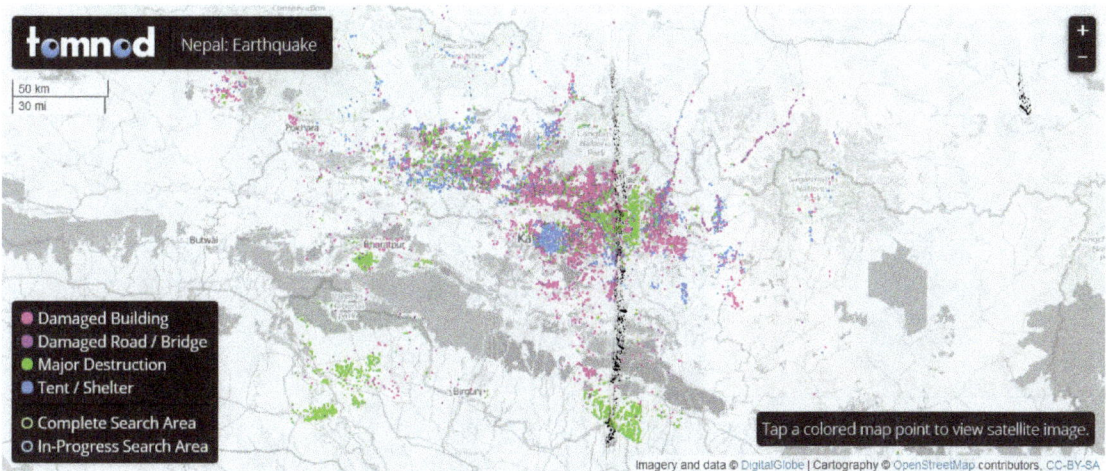

Figure 10. Earthquake in Nepal—Online damage map created by tomnod/OSM based on WorldView-1,-2,-3 and Geo-Eye [46]

pathfinding to get a better overview and understanding of problems on site. In addition, these maps proved to be very useful for making decisions on logistics and joint operations among relief organisations. Nevertheless, a multitude of websites and platforms hosted the maps, which resulted for some users in an overflow of mapping and imagery. During such extreme events, the authors of source [15] recommended a better coordination and harmonisation of global mapping efforts. As a consequence, an International Working Group on Satellite based Emergency Mapping (IWG-SEM) was initiated in 2012 resulting from the experiences of the Haiti earthquake in 2011. The IWG-SEM is a voluntary group of organisations involved in satellite-based emergency mapping which supports disaster response by improving international cooperation in such mapping activities. The group was founded to improve cooperation, communication and professional standards among the global network of satellite-based emergency mapping providers.

3.3. Forest fire in Russia

In July 2015, a heat wave in Russia's Siberian district started with over one hundred forest fires in the Buryatia and Irkustkaya Republics. Fires spread to an area of approximately 100,000 hectares [50].

After a request from the Russian Federal Space Agency (ROSCOSMOS) together with the Agency for Support and Coordination of Russian Participation in International Humanitarian Operations(EMERCOM), the Charter was triggered [Charter]. Burned area was mapped using the Russian satellite Resurs-P and its multispectral sensor with 12 meter resolution data (see Figure 10) and furthermore with SPOT-7 and GAOFEN-1 for another area [50]. In addition, fire hot spots were detected using RapidEye, GAOFEN and SPOT 7 [e.g. see Figure 11]. Burnt area analysis and fire hot spot detection was obtained by visual interpretation.

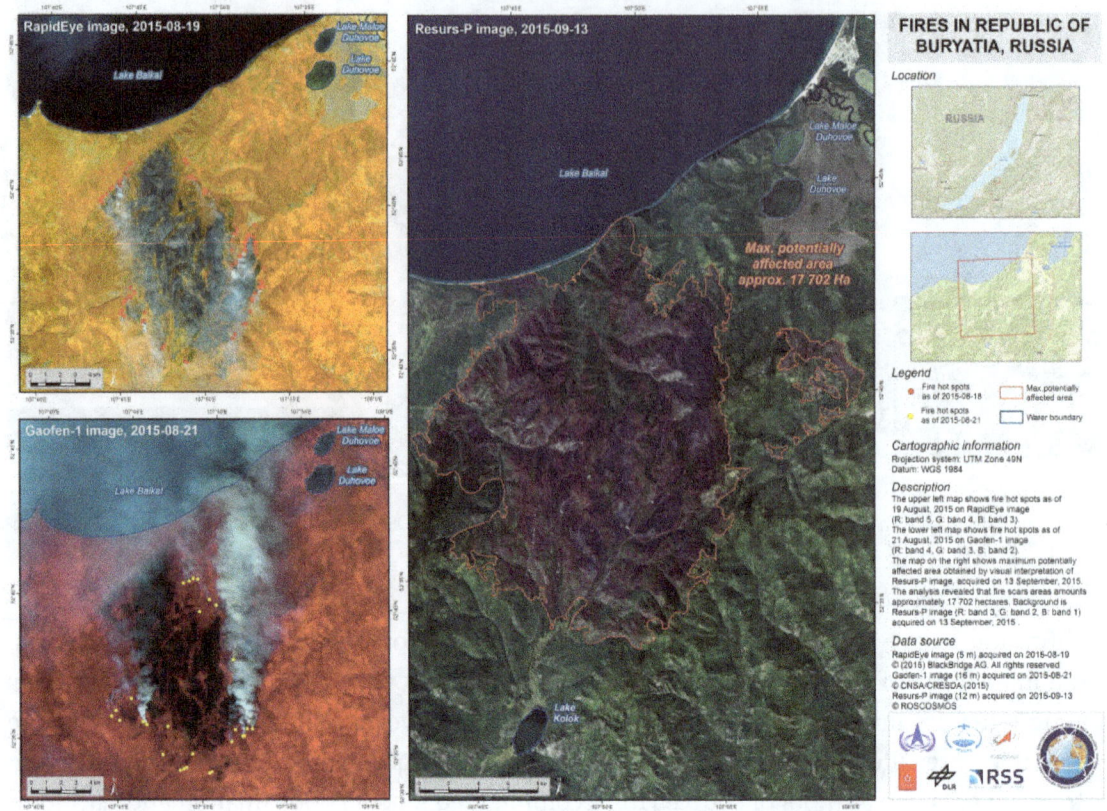

Figure 11. Forest Fires in Russia 2015 – Mapping hot spots and burned areas created by Research Center for Earth Operative Monitoring (NTs OMZ) [48]

In general, to identify fires and/or burnt areas optical data are the best choice and either semi-automated methods or visual interpretations can be applied [51]. In case of cloud coverage or direct fire effects such as smoke plumes or haze, visual interpretation usually gives better results. If the area is completely covered with clouds, radar data is an option. First positive results were gathered using SAR data for burnt area detection, applying backscatter coefficient analysis [52–53]. Nevertheless, optical satellites are commonly the main data source used for burnt area mapping in a rush-mode. In addition, fire hot spots can be detected automatically with optical data and appropriate methods. For instance, based on data of the NASA-owned MODIS sensors on board of the Terra-1 and Aqua-1 satellites, active fires can be detected [54].

3.4. Search for Malaysia Airlines Flight 370

On 11th March 2014, China Meteorological Administration requested the Charter for supporting search for Malaysia Airlines Flight 370 (MH370) [55]. The aircraft disappeared from radar on March 7th 2014 with 239 people on board. Soon an international search began in the South China Sea, the last known location of the aircraft. A few days later the search area has been expanded several times (e.g. to Indian Ocean). Satellite imagery was used to search for any evidence of the aircraft. Satellite images have revealed suspected debris or oil spill of missing aircraft in a number of locations (see example in Figure 12), but despite efforts to search an area of the southern Indian Ocean no trace of the plane has been found [55].

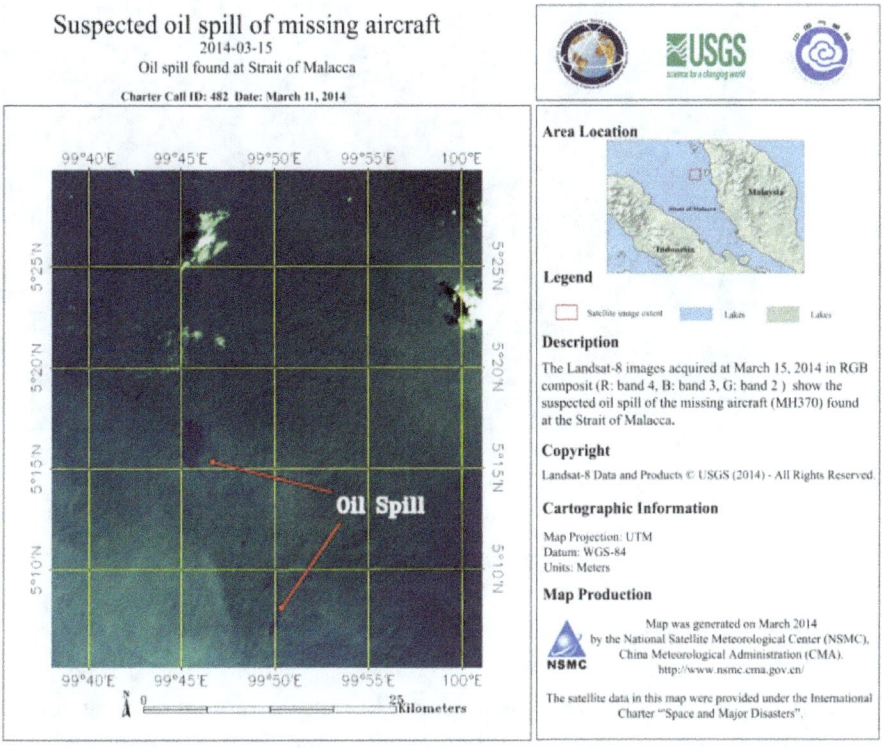

Figure 12. Searching for Malaysia Flight 370—Suspected oil spill of missing aircraft; created by National Satellite Meteorological Center (NSMC), China Meteorological Administration (CMA) [55]

Tomnod, a company owned by the satellite provider DigitalGlobe, started a crowdsourcing campaign in which over two million volunteers have studied WorldView-2 images of the area. The search area was sliced up in many small images which every user was able to see and tag with four types: Wreckage, Oil slick, Life raft and Other. Like other microtasking platforms, Tomnod uses triangulation to calculate areas of greatest consensus by the crowd [56]. The results are illustrated in Google Earth (see Figure 13).

Even though crowdsourced satellite information gave some reliable results or even misjudgements for this disaster, this way of data provision and processing information has been useful in man-made disasters before [56-57] and is a powerful new approach of producing crisis information based on satellite data. Nevertheless, Tomnod has to improve their infrastructure such as adding server capacity. Due to the huge volume of traffic (an estimated 100,000 views per minute), the site was down for several hours on March 11th and 12th.

However, the information provided by the Charter was also wrong. It is very difficult to search small items even if the imagery has a spatial resolution of 0.5 meters in an unspecific area with the current Earth observation technology. However, still there are a lot of data (e.g. SAR or other complex information) and several applications in disaster management which need advanced image analysis procedures. Therefore, remote sensing experts and specific tools are mandatory.

Figure 13. Searching for Malaysia Flight 370 – Online damage map created by tomnod based in Worldview-2 data and illustrated in Google Earth [56-57, visualisation in GoogleEarth]

4. Conclusion and outlook

The increasing occurrence of natural disasters and humanitarian emergency situations cause a growing demand for timely and up-to-date geoinformation for an effective disaster management. Within the last 10–15 years, a promising and considerable development has taken place to improve and accelerate the provision of Earth observation-based disaster information. Accordingly, remote sensing technology plays an important role in disaster management, especially during the preparedness and response phase.

Examples could be shown of how operational mechanisms (e.g. Charter, Copernicus EMS, ZKI-DE, etc.) serve rapid mapping based on Earth Observation data. In addition, a number of potential remote sensing data sources (TerraSAR-X, SPOT, RapidEye, WorldView and airborne data) and new (semi-automated) algorithms as well as visual interpretation results could be showcased for different disaster types.

Nevertheless, there are still some limitations with respect to the rapid availability of imagery and the reliability of the rapid image analysis in case of a disaster. Even if the imagery has a spatial resolution of 0.5 meters, geometric resolution is often still too coarse to assess damage or other important disaster information. Moreover, compromise must be found between the time spent on an analysis and the mapping accuracy that needs to be achieved. These aspects also have to be evaluated against user requirements during an emergency.

Further effort and scientific work is needed to derive even better, faster and more standardised crisis information from space-based imagery. In addition, in case of extreme disasters, a more structured and coordinated way of collaboration is needed to achieve most powerful results [15]. The International Working Group on Satellite based Emergency Mapping provides an important framework in this context.

Within the next years, new data sources (better geometric, spectral and temporal resolution), new tools, modified data policies (open access or strictly regulated) and new actors/collaborations (crowdsourcing crisis mapper, national services, inter-organisational cooperation) will influence the potential of remote sensing for natural or man-made disasters.

Author details

Monika Gähler*

Address all correspondence to: monika.gaehler@dlr.de

German Aerospace Center (DLR), Center for Satellite Based Crisis Information (ZKI) of the German Remote Sensing Data Center (DFD), Oberpfaffenhofen / Wessling, Germany

References

[1] Guha-Sapir D., Hargitt D., Hoyois P. Thirty Years of Natural Disasters 1974–2003: The Numbers. Louvain: Univ. Louvain Presses. 2004. 76 p.

[2] Munich Re NatCatSERVICE. World Map of Natural Disasters 2014, Annual Statistics [Internet]. 2015. Available from: http://www.munichre.com/en/reinsurance/business/non-life/natcatservice/annual-statistics/index.html [Accessed: 2015-11-24]

[3] Cavallo E. A., Noy I. The Economics of Natural Disasters: A Survey. IDB Working Paper no. 35 [Internet]. 2009. Available from: http://ideas.repec.org/p/idb/wpaper/4649.html [Accessed: 2015-11-24]

[4] Schneiderhan T., Gähler M., Kranz O., Voigt S. Insights into the Emergency Mapping Service within the GMES project SAFER – Highlights, main achievements and challenges. In: Proceedings of the ESA Living Planet Symposium, June 28–July 2, 2010, Bergen Norway. 2010. pp. 1–5

[5] Bello O. M., Aina Y. A. Satellite remote sensing as a tool in disaster management and sustainable development: Towards a synergistic approach. Procedia-Social and Behavioral Sciences. 2014. 120: 365–373. DOI:10.1016/j.sbspro.2014.02.114

[6] Van Westen C. Remote sensing for natural disaster management. International Archives of Photogrammetry and Remote Sensing. 2000. XXXIII: 1612–1617

[7] Centre for Research on the Epidemiology of Disasters (CRED): The OFDA/CRED International Disaster Database–Emergency Events Database (EM-DAT) [Internet]. 2015. Available from: http://www.emdat.be [Accessed: 2015-11-24]

[8] UNEP. IETC and Disaster Management: Looking at the Entire Cycle of Disaster Man-
     agement [Internet]. 2009. Available from: http://www.unep.or.jp/ietc/kms/data/
     2585.pdf [Accessed: 2015-11-24]

[9] Committee on Assessing the Costs of Natural Disasters, National Research Council,
     Division on Earth and Life Studies, Commission on Geosciences, Environment and
     Resources. The Impacts of Natural Disasters: A Framework for Loss Estimation. Na-
     tional Academies Press. Washington. 1999. 80p.

[10] Coppola D. Introduction to International Disaster Management. 3rd Edition. Butter-
     worth-Heinemann. Oxford. 2015. 760p.

[11] Campbell J. B. and Wynne R. H. Introduction to Remote Sensing, 5th Edition. The
     Guilford Press. New York. 2011. 662 p.

[12] Lewis S. Remote Sensing for Natural Disasters: Facts and Figures, Science and Devel-
     opment Network [Internet]. 2009. Available from: http://www.scidev.net/global/
     earth-science/feature/remote-sensing-for-natural-disasters-facts-and-figures.html
     [Accessed: 2015-11-24]

[13] Joyce K. E., Belliss S. E., Samsonov S. V., McNeill S. J., Glassey P. J. A review of the
     status of satellite remote sensing and image processing techniques for mapping natu-
     ral hazards and disasters. Progress in Physical Geography. 2009. 33(2): 183–207. DOI:
     10.1177/0309133309339563

[14] Van Westen C. J. Remote sensing and GIS for natural hazards assessment and disas-
     ter risk management. In: Treatise on Geomorphology. Schroder J. F., Bishop M. P.,
     editors. San Diego: Academic Press, Elsevier, 2013 (Remote Sensing and GIScience in
     Geomorphology, vol. 3). pp. 259–298.

[15] Voigt S., Schneiderhan T., Twele A., Gähler M., Stein E., Mehl H. Rapid damage as-
     sessment and situation mapping: Learning from the 2010 Haiti Earthquake. Photo-
     grammetric Engineering & Remote Sensing. 2011. 77(9): 923–931.

[16] Joyce K. E., Wright K. C., Samonsov S. V., Ambrosia V. G. Remote sensing and the
     disaster management cycle. Geoscience and Remote Sensing. Vienna: In-Tech Pub-
     lishing 2011: 317–346. DOI: 10.5772/8341

[17] European Commission, Joint Research Centre: European Forest Fire Information Sys-
     tem [Internet]. 2015. Available from: http://forest.jrc.ec.europa.eu/effis/about-effis
     [Accessed: 2015-11-24]

[18] Voigt S., Kemper T., Riedlinger T., Kiefl R., Scholte K., Mehl H. Satellite image analy-
     sis for disaster and crisis-management support. IEEE Transactions on Geoscience and
     Remote Sensing. 2007. 45(6): 1520–1528.

[19] Voigt S., Kranz O. Satellite-based crisis management and geoinformation systems.
     Safety & Security International. 2008. 4: 2–4

[20] Gähler M., ZKI Team. Support Disaster Management with Remote Sensing. In: Schiewe J., Michel U. (eds): Geoinformatics Paves the Highway to Digital Earth, Osnabrück. 2008. pp. 24–29

[21] European Space Agency (ESA). International Charter Space and Major Disasters. 14th Annual Report [Internet]. 2015. Available from: https://www.disasterscharter.org/documents/10180/66908/14thAnnualReport [Accessed: 2015-11-24]

[22] Kaku K., Held H. Sentinel Asia: A space-based disaster management support system in the Asia-Pacific region. International Journal of Disaster Risk Reduction. 2013. 6: 1–17. DOI: 10.1016/j.ijdrr.2013.08.004

[23] Sentinel Asia. [Internet]. 2015. Available from: https://sentinel.tksc.jaxa.jp/sentinel2/topControl.jsp [Accessed: 2015-11-24]

[24] European Commission: Copernicus Programme [Internet]. 2015. Available from: http://www.copernicus.eu [Accessed: 2015-11-24]

[25] European Commission: Copernicus Emergency Management Service [Internet]. 2015. Available from: http://emergency.copernicus.eu [Accessed: 2015-11-24]

[26] European Space Agency (ESA). Copernicus – Observing the Earth [Internet]. 2015. Available from: http://www.esa.int/Our_Activities/Observing_the_Earth/Copernicus/Overview3 [Accessed: 2015-11-24]

[27] Unitar's Operational Satellite Applications Programme – UNOSAT [Internet]. 2015. Available from: http://www.unitar.org/unosat/ [Accessed: 2015-11-24]

[28] UN-SPIDER Knowledge Portal – Emergencies Mechanisms [Internet]. 2015. Available from: http://www.un-spider.org/space-application/emergency-mechanisms [Accessed: 2015-11-24]

[29] NASA-SERVIR Global [Internet]. 2015. Available from: https://www.servirglobal.net [Accessed: 2015-11-24]

[30] DLR – ZKI-DE Service [Internet]. 2015. Available from: http://www.zki.dlr.de/services/zki-de [Accessed: 2015-11-24]

[31] Munich Re – Floods Dominate Natural Catastrophe Statistics in First Half of 2013 [Internet]. 2013. Available from: http://www.munichre.com/en/media-relations/publications/press-releases/2013/2013-07-09-press-release/index.html [Accessed: 2015-11-24]

[32] European Space Agency (ESA). International Charter Space and Major Disasters. Flooding in Germany [Internet]. 2013. Available from: https://www.disasterscharter.org/web/guest/-/floods-in-germany [Accessed: 2015-11-24]

[33] European Commission: Copernicus Emergency Management Service. Flooding in Germany [Internet]. 2013. Available from: http://emergency.copernicus.eu/mapping/list-of-components/EMSR044 and EMSR047 [Accessed: 2015-11-24]

[34] DLR – ZKI-DE. Flooding in Germany [Internet]. 2013. Available from: http://
      www.zki.dlr.de/article/2373 [Accessed: 2015-11-24]

[35] LAWA, Bund/Länder-Arbeitsgemeinschaft Wasser. Zusammenfassende Analyse der
      Ergebnisse der vom Hochwasser 2013 betroffenen Flussgebietsgemeinschaften [Inter-
      net]. 2014. Available from: http://www.lawa.de/documents/LAWA_Analyseber-
      icht_Hochwasser_Juni_2013_dae.pdf [Accessed: 2015-11-24]

[36] Martinis S., Kersten J., Twele A. A fully automated Terra SAR-X based flood service.
      ISPRS Journal of Photogrammetry and Remote Sensing. 2015. 104: 203–212. DOI:
      10.1016/j.isprsjprs.2014.07.014

[37] Martinis S, Twele A, Strobl C, Kersten J, Stein E. A multi-scale flood monitoring sys-
      tem based on fully automatic MODIS and TerraSAR-X processing chains. Remote
      Sensing. 2013. 5: 5598–5619. DOI: 10.3390/rs5115598

[38] Martinis S., Künzer C, Wendleder A, Huth J, Twele A, Roth A, Dech S. Comparing
      four operational SAR-based water and flood detection approaches. International
      Journal   of   Remote   Sensing.   2015.   36(13):   3519–3543.   DOI:
      10.1080/01431161.2015.1060647.

[39] Pulvirenti L., Pierdicca N., Chini M., Guerriero L. An algorithm for operational flood
      mapping from Synthetic Aperture Radar (SAR) data using fuzzy logic. Natural Haz-
      ards Earth System Sciences 2011. 11: 529–540. DOI:10.5194/nhess-11-529-2011

[40] European Commission: ECHO Fact Sheet – Nepal [Internet]. 2015. Available from:
      http://ec.europa.eu/echo/files/aid/countries/factsheets/nepal_en.pdf   [Accessed:
      2015-11-24]

[41] European Commission: Copernicus Emergency Management Service. Earthquake in
      Nepal [Internet]. 2015. Available from: http://emergency.copernicus.eu/mapping/list-
      of-components/EMSR125 [Accessed: 2015-11-24]

[42] European Space Agency (ESA). International Charter Space and Major Disasters.
      Earthquake in Nepal [Internet]. 2015. Available from: https://www.disasterschar-
      ter.org/web/guest/-/landslide-in-nep-2 [Accessed: 2015-11-24]

[43] UN-SPIEDER Knowledge Portal – Earthquake in Nepal, India, Bangladesh, China
      [Internet]. 2015. Available from: http://www.un-spider.org/advisory-support/emer-
      gency-support/9911/earthquake-nepal-india-bangladesh-china   [Accessed:
      2015-11-24]

[44] DLR – ZKI-DE. Earthquake in Nepal [Internet]. 2015. Available from: http://
      www.zki.dlr.de/article/2748 [Accessed: 2015-11-24]

[45] iRevolutions. A Force for Good: How Digital Jedis are Responding to the Nepal
      Earthquake [Internet]. 2015. Available from: http://irevolution.net/2015/04/27/digital-
      jedis-nepal-earthquake [Accessed: 2015-11-24]

[46] Tomnod. Nepal Earthquake Data Portal [Internet]. 2015. Available from: http://blog.tomnod.com/Nepal-Earthquake-Data-Portal/ [Accessed: 2015-11-24]

[47] Stramondo S., Bignami C., Chini M., Pierdicca N., Tertulliani A. The radar and optical remote sensing for damage detection: Results from different case studies. International Journal of Remote Sensing. 2006. 27: 4433–4447

[48] Gamba P., Dell'Acqua F., Trianni G. Rapid damage detection in Bam area using multi-temporal SAR and exploiting ancillary data. IEEE Transactions on Geoscience and Remote Sensing. 2007, 45(6): 1582–1589

[49] Chini M., Pierdicca N., Emery W. J. Exploiting SAR and VHR optical images to quantify damage caused by the 2003 Bam earthquake. IEEE Transaction on Geosciences and Remote Sensing. 2009. 47(1): 145–152

[50] European Space Agency (ESA). International Charter Space and Major Disasters. Fire Russia [Internet]. 2015. Available from: https://www.disasterscharter.org/web/guest/-/fire-in-russian-federation [Accessed: 2015-11-24]

[51] Liew S. C., Kwoh L. K., Lim O. K., Lim H. Remote sensing of fire and haze. In: Eaton P., Radojevic M., editors. Forest Fires and Regional Haze in Southeast Asia. New York: Nova Science Publishers, 2001. pp. 67–89

[52] Tanase M. A., Santoro M., Wegmüller U., de la Riva J., Pérez-Cabello F. Properties of X-, C-and L-band repeat-pass interferometric SAR coherence in Mediterranean pine forests affected by fires. Remote Sensing of Environment. 2010. 114(10): 2182–2194. DOI: 10.1016/j.rse.2010.04.021

[53] Bernhard E. M., Twele A., Martinis S. The effect of vegetation type and density on X-band SAR backscatter after forest fires. Photogrammetrie, Fernerkundung, Geoinformation. 2014. 4: 275–285. DOI: 10.1127/1432-8364/2014/0222

[54] Giglio L., Descloitresa J., Justicec C. O., Kaufman Y. J. An enhanced contextual fire detection algorithm for MODIS. Remote Sensing of Environment. 2003. 87(2/3): 273–282.

[55] European Space Agency (ESA). International Charter Space and Major Disasters. Missing Malaysia Airlines Jet [Internet]. 2015. Available from: https://www.disasterscharter.org/web/guest/-/missing-malaysia-airlines-jet [Accessed: 2015-11-24]

[56] Tomnod. Search for Flight MH370 [Internet]. 2015. Available from: http://www.tomnod.com/campaign/malaysiaairsar2014/map/15qx1ry59 [Accessed: 2015-11-24]

[57] iRevolutions. Results of the Crowdsourced Search for Malaysia Flight 370 [Internet]. 2015. Available from: http://irevolution.net/2015/04/27/digital-jedis-nepal-earthquake http://irevolution.net/2014/03/15/results-of-the-crowdsourced-flight-370-search/ [Accessed: 2015-11-24]

7

Remote Sensing of the Ocean Environment Using Finite Element Methods

Saba Mudaliar, C.P. Vendhan and C. Prabavathi

Additional information is available at the end of the chapter

Abstract

Oceans are a vast, complex world where underwater sound is the most efficient tool available to understand its detailed characteristics. However the underwater channel has a very complex geometrical and material structure and hence special techniques are required to model it. Analytical solutions are feasible only when one makes gross assumptions and approximations. Several numerical and semi-numerical techniques have been developed for estimating the sound field in the ocean channel. But no single method is capable of handling all possible environmental conditions, frequency, and ranges of interest in remote sensing problems. We explore in this chapter the scope and feasibility of finite element method in underwater remote sensing. The current study is based on a channel model with cylindrical symmetry and a time-harmonic source signal. A variational formulation is used to derive the finite element model for acoustical radiation, scattering and propagation in the ocean. A Bayliss-type radiation boundary condition is used to model the far field behaviour without the need to deal with a large solution domain. Since the ocean geometry can support several propagating, evanescent, and radiation modes, a penalty function approach is employed to impose the far field radiation condition. A distinct feature of the ocean channel is its depth-dependent sound speed. The eigensolution for this channel is required for imposing the radiation condition at the truncation boundary. We have cast this eigenproblem in a variational form and employed a Rayleigh-Ritz method to obtain an approximate eigensolution. This approach has provided a good approximation of the depth eigenmodes in a compact semi-analytic form. We have employed our finite element algorithm to model several range- and depth-dependent ocean problems. Our numerical study has established that our finite element algorithm gives accurate results with reasonable effort. In particular, our finite element approach is most appropriate for shallow water problems where the interaction of wave modes with irregular ocean bottom is quite complex. The penalty function approach employed to implement the radiation boundary condition has been found to be robust over a wide range of penalty scale factors. We have also extended this work for the case of irregular elastic sea bed. We continue to explore and further develop our finite element approach by applying it to several other ocean acoustic problems encountered in the remote sensing of ocean environment.

Keywords: Wave propagation, scattering, ocean wave guide, irregular boundaries

---

1. Introduction

Oceans are a vast, complex, mostly dark, optically opaque but acoustically transparent world which is only thinly sampled by today's limited science and technology. Underwater sound[1] is used as the premier tool to determine the detailed characteristics of physical and biological bodies and processes in the ocean. The distributions within the sea of the physical variables affect the transmission of sound. The wide range of acoustic frequencies and wavelengths, together with the diverse oceanographic phenomena that occur over full spectra of space and time scales, thus give rise to a number of interesting effects and opportunities. Because of its great practical importance, especially to naval submarine operations, ocean-acoustics research [1-5] has been driven by applications more than other branches of ocean science.

Acoustic remote sensing in a generic sense refers to sending out acoustic signals and recording the scattered waves, which is hence processed to ascertain the nature of target/obstruction that was encountered by the transmitted signal. This remote sensing in general involves transmission, processing of received signals and some form of inversion. This chapter is exclusively dedicated to accurate modeling of propagation and scattering of acoustic signals in the ocean channel.

The amplitude and phase of sound field generated by an acoustic source in the ocean can be deduced, in principle, by solving either the wave equation or the Helmholtz equation in the case of a harmonic acoustic source [1]. However, this procedure is generally difficult to implement due to the complexity of the ocean-acoustic environment: the sound-speed profile is usually non-uniform in depth and/or range, giving rise to waveguide focusing and shadowing effects; the sea surface is rough and time-dependent; the ocean floor is typically a very complex, rough boundary which may be inclined to the horizontal; and the bottom may be an elastic medium, capable of supporting shear waves along the ocean-bottom boundary. To compound the problem, various ocean processes, including internal waves and small-scale turbulence, introduce small fluctuations in the sound speed, which are responsible for significant acoustic fluctuations over long transmission paths.

Analytical solutions of the governing differential equations in underwater acoustics are not always feasible and can only be obtained if the sound speed of the water column and physical boundaries can be described in simple mathematical terms. This is rarely the case in reality and so it is generally necessary to employ approximate models. A variety of numerical

---

1 There exists a vast body of literature in remote sensing of ocean using electromagnetic and optical sensors from satellites. Although such methods have definite advantages in several aspects, they have serious limitations for sensing deep underwater channels.

techniques have been developed for estimating sound fields in the ocean, but no single method is capable of handling all possible environmental conditions, frequencies, and transmission ranges of interest in the applications. Even the existing ocean-acoustic propagation models [6-7] with restricted scope often take several hours to run on a supercomputer.

Several different approaches for the solution of the sound field in the ocean have evolved over the past few decades: ray tracing [8], normal-mode techniques [9] and coupled-mode models [10], the parabolic-equation approximation [11] and fast field programs (FFP) [12]. In this chapter, we will discuss in detail the scope of the finite element method [13-15] in ocean remote sensing applications. In order to motivate our finite element approach and put it in proper context, we briefly summarize the analytical and computational tools that are currently in use in the ocean remote sensing literature especially to point out their main merits and shortcomings.

2. Background

Ray-based methods [8-9] involve following the paths of a set of rays as they leave the source and tracking them as they propagate through the medium. They can be used for range-dependent and range-independent problems, but are most commonly used for range-independent problems. They are most useful for short-range, high-frequency modeling. Straightforward ray theory suffers from following drawbacks: (i) Need to deal with situations involving caustics and singularities. (ii) At each incidence on surface or bottom, each ray has to be "told" at what angle to go off, and with what percentage of total reflection. (iii) Since problems are almost entirely numerical, each variation is nearly as hard as the first try, e.g., a new source depth or a greater range. The main shortcoming of the ray method is the inherent high-frequency approximation.

A class of propagation models exist which gives the full-wave solution for the field in a horizontally stratified medium. This type of a model is known as "fast field program". This technique is basically a numerical implementation of the integral transform technique for horizontally stratified media [12, 9]. The field solution is in the form of a wavenumber integral which is evaluated by numerical quadrature. This approach is distinguished by its use of the fast Fourier transform (FFT) to calculate the integral. FFPs determine the field which satisfies the Helmholtz equation or similar equations which include shear wave effects. The Helmholtz equation for the stratified medium is a partial differential equation in two independent variables, range and depth, and hence in principle could be solved by the application of two integral transforms [12]. For certain specific sound-speed profiles having a particular analytical form, this can be achieved, yielding an exact solution for the field. For a general sound-speed profile, however, the transform over depth is intractable and an alternative technique must be sought. Nevertheless, a transform over range can be applied, and this is the starting point of the FFP argument [16]. In contrast to the ray solution, the FFP model yields a result which is essentially exact. Starting from the Helmholtz equation for a stratified medium, the only additional approximation is that of using the asymptotic approximation to the Bessel function.

This approximation turns out to include negligible errors beyond a wavelength or so from the source.

As an alternative to "exact" numerical propagation models, with their heavy computational overhead, a number of methods have been developed whose starting point is a parabolic equation [11]. Such an equation which is an approximation for the elliptic Helmholtz equation is valid over a small range of angles, usually, but not necessarily, extending about the horizontal. Given their inherently approximate nature, the parabolic-equation (PE) models are distinguished by a lack of precision, the extent of which depends by and large on the problem under consideration. They have acquired popularity amongst the ocean-acoustics community because they give the field over the entire water column with no additional effort and they can handle range-dependent environments. PE methods are often said to be valid within a cone of angles extending +/− 20° (narrow angle) and +/− 40° (wide angle) about the horizontal. One of the shortcomings of the PE models is that, when these angles are exceeded, the output continues to look reasonable, showing no obvious indication of error [7]. Apart from the excessive inaccuracy of these results, the lack of consistency among the PE codes highlights the general difficulty of assessing their performance in any given environment. Although the PE is relatively easy to implement, there is a price to be paid: a) it is valid over only a limited range of angles, a consequence of the paraxial approximation, and b) it is a one-way solution, capable of handling only outgoing waves, since incoming radiation, represented by a Hankel function of the second kind of zero order, is neglected in the solution. Little can be done to remedy the backscatter limitation, but considerable effort has gone into extending the angular range of the forward-scatter regime [17]. The advantage of the parabolic equation over the original Helmholtz equation is that the PE can be solved by a straightforward marching in range which requires much less computational effort. From a numerical point of view, this range marching is typically implemented using either standard finite difference techniques or using a fast Fourier transform as in the so-called split-step method. There are other approaches to solve the parabolic wave equation in ocean waveguides. Lee et al. [18] employed the finite difference method whereas Huang [19] used a finite element method to solve the PE.

The sound field in a horizontally stratified ocean can be expressed as an infinite sum of uncoupled normal modes plus one or more branch line integrals [9, 1]. At large ranges from sources, the branch line integral component is negligible and the field is given accurately by the normal-mode sum, but in the vicinity of the source, within the cycle distance of each mode, the integrals are significant and should be taken into account. If the environment shows some range dependence, through either the sound-speed profile or the boundary conditions, the field is no longer separable and strictly (uncoupled) normal-mode theory does not apply. However, provided the range dependence is sufficiently slow, the adiabatic approximation is valid, i.e. there is essentially no transfer of energy between modes as they propagate the channel. If the range dependence is too fast for the adiabatic approximation to hold, mode coupling is significant, which requires the calculation of the coupling coefficients—a time consuming procedure. The normal-mode method is typically accurate for ranges greater than the first 10 water depths or so, a figure which depends on the number of modes that are included in the solution. In the near field, more modes should be computed to closely predict the fields accurately. The normal-mode models tend to be thought of as providing solutions to range-independent problems. Range-dependent solutions can be obtained using (a) adiabatic mode theory or (b) coupled-mode theory. The later approach involves more computational cost but can provide more accurate results.

When the range dependence is too strong for mode coupling to be neglected, a different approach than the usual normal-mode theory is required. A complete two-way (i.e. including backscattering) solution to this problem has been formulated in terms of stepwise coupled normal modes [10]. The medium is sub-divided into a large number of thin vertical segments, in each of which the acoustic parameters are held constant in the range direction but are allowed to vary in depth. Across the segment boundaries, the pressure and horizontal particle velocity are required to be continuous. In this method, the field is expressed as a sum of local modes representing both outgoing and incoming cylindrical waves. Again, the modal eigenvalue problem has to be solved, and in this case, the Galerkin method is used, whereby the solution is expanded in a set of basis modes, yielding a tractable eigenvalue matrix problem [20]. This involves rather, complex coupling integrals which have to be evaluated for all modes at all segment boundaries. This method is computationally demanding, but it is essentially exact and forms the basis of the model [10, 21-22]. When the coupling effects are neglected, the full coupled-mode expressions reduce to the adiabatic approximation.

Ray tracing, normal-mode techniques, and coupled-mode models are accurate but computationally intensive; the parabolic equation is an approximation to the wave equation that has been solved using explicit and implicit finite difference schemes; Green's function solutions (fast field programs) are essentially models for which exact solutions are available that cannot account for sound-speed variation. If the variation of sound-speed profile is independent of range, the ocean is said to be horizontally stratified. Several of the numerical ocean-acoustic propagation models assume horizontal stratification. The advantage, from the point of view of the computation, is that the solution field separates into range and depth components, which simplifies the calculation of the field considerably. The speed of sound in the ocean shows only small departures from 1500 m/s, but nevertheless its effect on sound propagation on the ocean is profound. In the deep ocean, for example, the profile acts as an acoustic waveguide, supporting propagation to long ranges with little attenuation. However, for a general ocean environment, which has a range-dependent sound-speed profile, an ocean bed having irregular geometry, and turbulence in the water column, none of the existing methods described above work satisfactorily.

3. Finite element method

For general ocean environments, the finite element method (FEM) [13-15] is a good choice for the numerical modeling of ocean-acoustic wave propagation because it is exact within the limits of numerical accuracy and can accurately account for all the scattering processes. Although the literature on finite element technique on wave scattering and propagation is extensive, the number of available FEM models for ocean remote sensing is fairly small [23-24].

Part of the reason for this is the large computational cost involved. However, we feel that for shallow-water applications, the FEM is both feasible and appropriate. The very nature of the waves to radiate into the far field when unbounded requires the domain to be truncated with an artificial boundary, on which an approximate radiation boundary condition should be imposed [25-29]. In the present work, a variational approach is used to derive the finite element approximation for time-harmonic acoustic wave propagation in an axisymmetric, heterogeneous oceanic waveguide, and a BGT-type boundary damper [26] is used to model the effect of the far field. Since a waveguide in general supports multiple propagating/radiation modes in the far field, a penalty function approach has been employed to impose the modal radiation boundary condition in conjunction with the orthogonality property of the depth modes of the waveguide.

In our finite element model for depth- and range-dependent waveguides, the eigensolution of the depth problem is required for the imposition of the radiation condition at the truncation boundary. Unfortunately, the depth eigenproblem could be solved exactly only for a few special profiles. In view of this, several numerical methods have been developed to solve the depth problem [1]. Porter and Reiss [30] employ a finite difference model for the depth equation, and the resulting algebraic eigenproblem has been solved using a combination of iterative techniques and Richardson extrapolation to obtain the radial wavenumbers and modal vectors to a great degree of precision. For our finite element model [31-32], it would be convenient to have the depth modes in a compact analytical form. We have accomplished this by adopting the following procedure: The depth eigenproblem is cast in a variational form by suitably defining a functional. The classical Rayleigh–Ritz (RR) method is employed to find a variational approximation to the eigensolution of the depth problem in ocean-acoustic waveguides. The depth modes thus obtained have a compact semi-analytical form in contrast to methods using finite difference or other finite element methods. An interesting feature of the model is that the trial functions are derived from an isovelocity problem that has an exact solution. It is important to note that such trial functions automatically satisfy even the dynamic interface condition at the seabed, thus contributing to the accuracy of the numerical model. Our procedure has been tested for several different ocean profiles and the results compare well against those obtained using the method of Porter and Reis [33]. The proposed model thus provides an accurate representation of the depth eigenmodes in a compact semi-analytical form.

We have chosen several isovelocity waveguide examples, for which analytical solutions are available, to validate the FE model developed and ascertain its versatility to impose modal radiation boundary condition. We have confirmed the efficacy of the FE model by applying it to several examples of depth- and range-dependent waveguides. This numerical study establishes that our FE model gives accurate results with reasonable computational effort. The penalty function approach employed to implement the radiation boundary condition has been found to be robust over a wide range of penalty scale factors. We have also extended this work for the case of irregular elastic seabed. We continue to explore and further develop our FE model by applying it to several other ocean-acoustic problems encountered in the remote sensing of ocean environment.

4. Governing equations and boundary conditions

The fluid domain $\Omega = \Omega_I + \Omega_O$ (Fig. 1) of the waveguide problem consists of the inner domain Ω_I truncated by the artificial radiation boundary S_R, and the outer domain Ω_O (far-field domain). The waveguide is assumed to be axially symmetric about the vertical axis containing a source at depth z_s, with r denoting the radial coordinate or the range. It is bounded at the top by the $z = 0$ plane, which is the air–sea interface (S_F), and at the bottom by a seabed of arbitrary topography (S_B). The waveguide is assumed to have unbounded range. For time-harmonic linear acoustic waves with the pressure field denoted as $\hat{p}(r, z, t) = p(r, z)e^{-i\omega t}$, ω being the circular frequency of the source, the governing equation is given by

$$\rho\nabla\left(\tfrac{1}{\rho}\nabla p\right) + k^2 p = -\tfrac{1}{2\pi r} f_o \delta(r)\delta\left(z - z_s\right),\tag{1}$$

where ∇ is the gradient operator, ρ the density of the acoustic fluid, k the acoustic wavenumber, c the local speed of sound, and f_o defines the point source at $r = 0$ and $z = z_s$.

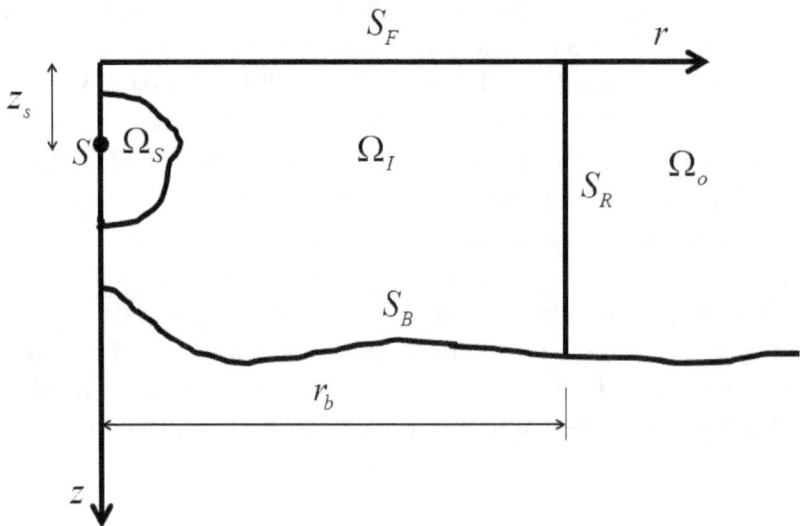

Figure 1. Geometry of the ocean waveguide

Considering the large impedance mismatch between air and water, a pressure release boundary condition may be used at the free surface. Thus,

$$p = 0 \qquad \text{on } S_F\tag{2}$$

As the waves encounter the seabed, there is partial reflection and the remaining energy is transmitted into the seabed. A part of the transmitted waves may be coupled back into the water column because of refraction through the sediment layers. However, for now, a rigid bottom is assumed, for which the normal derivative of the pressure should vanish at the bottom boundary. In other words,

$$\frac{\partial p}{\partial n} = 0 \quad \text{on } S_B, \tag{3}$$

where S_B denotes the sea bottom.

For the purpose of FE modeling, the waveguide, which is unbounded in range, is truncated at $r = r_b$, and the truncation boundary is treated as the radiation boundary S_R, on which a suitable approximate radiation condition should be imposed. Here, the boundary damper approach [26] has been adopted. The first-order cylindrical damper equation may be written as

$$\frac{\partial p_m}{\partial n} + \alpha p_m = 0, \qquad m = 1, 2, \cdots, M \quad \text{on } S_R, \tag{4}$$

where M denotes the number of propagating modes, and the damper coefficient α_m associated with the m-th mode is given by

$$\alpha_m = \frac{1}{2r} - ik_{rm}, \tag{5}$$

where k_{rm} denotes a horizontal wavenumber. It may be noted that Eq. (5) is exact for the asymptotic form of a cylindrically symmetric wave. On the truncation boundary S_R, acoustic pressure may be expressed as a sum of normal pressure modes as

$$p(z) = \sum_{m=1}^{M} p_m(z), \tag{6}$$

where $p(z)$, $m=1,2,...$, are the normal modes of propagation for the problem in Eq. (1). Following Fix and Marin [31], the radiation boundary condition for the waveguide problem may be written, using Eqs. (4) and (6), as

$$\frac{\partial p}{\partial r} + \sum_{m=1}^{M} \alpha_m p_m \quad \text{on } S_R. \tag{7}$$

Denoting a normal-mode function at the radiation boundary by $f_m(z)$, which is associated with the m-th propagating mode eigenvalue, the pressure modes in Eq. (6) may be written as

$$p_m(z) = a_m f_m(z), \qquad m = 1, 2, \cdots, M, \tag{8}$$

where a_m denotes a modal participation factor. Then the radiation boundary condition in Eq. (7) may be rewritten as

$$p(z) + \sum_{m=1}^{M} a_m \alpha_m f_m(z) = 0 \qquad \text{on } S_R, \tag{9}$$

where the constants a_m are determined by using the $(1/\rho(z))$ -orthogonality of the normal modes. It has tacitly been assumed here that the waveguide has constant water depth and range-independent but depth-dependent sound speed in the vicinity of the truncation boundary S_R and beyond, so that the depth eigenproblem corresponding to the problem in Eq. (1) could be solved at least numerically [23].

Note that while the radiation condition in Eq. (4) on an individual mode is local, the radiation condition in Eq. (9) is global, meaning that nodes of an element on the truncation boundary are linked to other elements there in view of the coefficients a_m.

5. Constraints

In view of Eq. (8), Eq. (6) may be written as

$$[C]\{p(z), a_1, a_2, \ldots, a_M\}^T = 0 \quad \text{on } S_R, \tag{10}$$

where

$$[C] = \left[1, -f_1(z), -f_2(z), \ldots, -f_M(z)\right], \tag{11}$$

and the companion vector in Eq. (10) is unknown. Equation (10) will be treated as a constraint in the following FE model.

6. Variational formulation

For the purpose of finite element modeling, it would be convenient to construct a variational formulation [34-35]. In the present study, in order to avoid possible numerical difficulties with handling a point source, a small fluid domain Ω_S surrounding the source has been excluded so that the computational domain is $\bar{\Omega}_I = \Omega_I - \Omega_S$. Consider the following axisymmetric functional $I(p)$ defined in the cylindrical coordinate system (r, z) (see Fig. 1):

$$I(p) = \frac{1}{2} \int_{\bar{\Omega}_I} \frac{1}{\rho} \left\{ \left(\frac{\partial p}{\partial r} \right)^2 + \left(\frac{\partial p}{\partial z} \right)^2 - k^2 p^2 \right\} r \, dr \, dz + \frac{1}{2} \sum_{m=1}^{M} \int_{S_R} \frac{1}{\rho} \alpha_m p_m^2 \, r \, dz - \int_{S_D + S_N} \frac{1}{\rho} \frac{\partial p}{\partial n} p \, dS, \tag{12}$$

where S_D denotes the surface on which a Dirichlet boundary condition is prescribed and S_N the surface with prescribed Neumann boundary condition, and the other domains of integration are identified in Fig. 1.

It can readily be shown that the variational condition

$$\delta I = 0 \tag{13}$$

leads to the governing differential equation in Eq. (1) and the boundary conditions in Eqs. (2)–(4). Thus, Eqs. (12) and (13) can be used to develop an FE model using the Rayleigh–Ritz approximation. However, the resulting solution should also obey the constraints in Eq. (10), which will ensure the imposition of the radiation boundary condition as discussed above. This will be achieved by modifying the discrete approximation to the functional in Eq. (12).

7. Finite element model

The finite fluid domain $\bar{\Omega}_I$ (which excludes the source) of the axisymmetric waveguide in Fig. 1 may be discretized using eight-noded axisymmetric quadrilateral elements with C_o-continuity and the well-known isoparametric formulation [15]. The computational domain is discretized into a mesh of finite elements. The finite element approximation for the field variable p may then be written as

$$p(r,z) \approx \sum_{j=1}^{\tilde{n}} \bar{p}_{ej} N_j(\xi, \zeta) = \left[N \right]^T \{ \bar{p}_e \}, \tag{14}$$

where $\tilde{n}$ denotes the number of element nodes (eight in the present study), $\bar{p}_{ej}$ the nodal pressure variable/degrees of freedom (dofs) and $N_j(\xi, \zeta)$ the polynomial shape function in the parametric coordinates (ξ, ζ) in the (r, z) plane [15]. The subscript e is used to indicate the quantity at the element. Substituting Eq. (14) into Eq. (12) yields the following discrete form:

$$I(p_e) \approx \frac{1}{2} \{ \bar{p}_e \}^T \left(\left[K_e \right] - \left[M_e \right] \right) \{ \bar{p}_e \} + \frac{1}{2} \sum_{m=1}^{M} \{ \bar{p}_{em} \}^T \left[R_{em} \right] \{ \bar{p}_{em} \} - \{ \bar{p}_e \}^T \{ f_e \}, \tag{15}$$

where $\{ \bar{p}_{em} \}$ denotes nodal pressure on the radiation boundary due to the m-th mode and the various matrices above will be identified subsequently. The stationary condition of the potential $I(p_e)$ above should be sought subject to the constraint in Eq. (10). There are two ways of implementing this, one the classical Lagrangian multiplier approach and the other the penalty function approach; the latter, which is commonly used in the context of finite element analysis [15, 13] is adopted in the present work. To achieve this, a modified potential I' may be defined as

$$I' = I + \tfrac{1}{2}\{\bar{p}'_e\}^T [C_e]^T [\beta_P][C_e]\{\bar{p}'_e\}, \tag{16}$$

where $[C_e]$ denotes the constraint matrix in Eq. (11) specific to an element. The penalty coefficient matrix $[\beta_P]$ above may be chosen to be diagonal for convenience, with β_{Pm} denoting the penalty parameter associated with the m-th mode. Equation (16) may be expanded as

$$I'(p'_e) = \tfrac{1}{2}\{\bar{p}'_e\}^T \left([K'_e] - [M'_e] + [R'_e] + [C_e]^T [\beta_P][C_e]\right)\{\bar{p}'_e\} - \{\bar{p}'_e\}^T \{f'_e\},$$

where the enlarged element dof vector is defined as

$$\{\bar{p}'_e\} = \left\{ \begin{matrix} \{\bar{p}_e\} \\ \{a\} \end{matrix} \right\}. \tag{17}$$

The enlarged stiffness, mass and damping matrices, and load vector in expanded Eq. (16), consistent with $\{\bar{p}'_e\}$ in Eq. (17), are given by

$$[K'_e] = \begin{bmatrix} [K_e] & 0 \\ 0 & 0 \end{bmatrix} \quad [M'_e] = \begin{bmatrix} [M_e] & 0 \\ 0 & 0 \end{bmatrix} \quad [R'_e] = \begin{bmatrix} 0 & 0 \\ 0 & [R_e] \end{bmatrix} \quad \{f'_e\} = \left\{ \begin{matrix} \{f_e\} \\ 0 \end{matrix} \right\}. \tag{18}$$

The matrices $[K_e]$, $[M_e]$, and $[R_e]$ in Eq. (18) are traditionally called the element stiffness, mass and radiation damping matrices, and $\{f_e\}$ the load vector, respectively. They are given as follows:

$$[K_e] = \int_{\Omega_e} \tfrac{1}{\rho}[\nabla N]^T [\nabla N]\, d\Omega \tag{a}$$

$$[M_e] = \int_{\Omega_e} \tfrac{k^2}{\rho}[N]^T [N]\, d\Omega \tag{b}$$

$$[R_e] = \mathrm{diag}[R_{e1}, R_{e2}, \ldots, R_{eM}] \tag{c}$$

$$R_{em} = \{f_{zm}\}^T [\bar{R}_{em}]\{f_{zm}\} \qquad [\bar{R}_{em}] = \alpha_m \int_{S_{Re}} \tfrac{1}{\rho}[N]^T [N]\, dS \tag{d}$$

$$\{f_{zm}\} = \left\langle f_{zm}(z_1), f_{zm}(z_2), \ldots, f_{zm}(z_n)\right\rangle^T \tag{e}$$

$$\{f_e\} = \int_{S_{Ne}} \tfrac{1}{\rho} p_v [N]^T\, dS \tag{f}$$

$$\tag{19}$$

where $[N]$ denotes the shape-function matrix [see Eq. (14)], $p_v = \partial p / \partial n$, and $f_{zm}(z_j)$ denotes the j-th nodal value of the m-th mode on a finite element in contact with the radiation boundary S_{Re}. The steps required to derive Eq. (19c) are outlined in Appendix A. It is of interest to note that the radiation-damping matrix $[R_e]$ in Eq. (19c) implies uncoupled modal participation. However, the constraint term involving the matrix $[C_e]$ in expanded Eq. (16) brings about modal coupling. The various integrals above are defined over relevant finite element domains. The stationary condition of the potential I' in expanded Eq. (16) is obtained by setting

$$\frac{\partial I'}{\partial \{\overline{p}_e'\}} = 0 \tag{20}$$

Equation (20) leads to general element equations of the form

$$\left([K_e'] - [M_e'] + [R_e'] + [C_e]^T [\beta_P][C_e]\right)\{\overline{p}_e'\} = \{f_e'\}. \tag{21}$$

It may be noted that if the penalty matrix $[\beta_P] = 0$ in Eq. (21), the constraints are ignored; as the penalty parameter values increase, the error in satisfying the constraint equations decreases, and for very high values of penalty, the numerical solution may break down. Hence, a judicious choice of the penalty parameters is essential. The radiation-damping matrix $[R'_e]$ and the constraint matrix $[C_e]$ in Eq. (21) correspond to elements on the radiation boundary. Hence, for this case, the FE equation may be deduced from Eqs. (18) and (21) as

$$\left([K_e] - [M_e]\right)\{\overline{p}_e\} = \{f_e\}. \tag{22}$$

The radiation-damping matrix $\lfloor R'_e\rfloor$ in Eq. (19c), which is complex in view of Eq. (5), is defined only for elements that share one or more of their boundaries with the artificial boundary S_R. Carrying out thus the finite element assemblage operation, yields the following global finite element equations:

$$\left([K'] - [M'] + [R'] + [C']\right)\{\overline{p}'\} = \{f'\}, \tag{23}$$

where the global solution vector $\{\overline{p}'\}$ consists of all the pressure dof in the computational domain as well as the unknown vector $\{a\}$ in Eq. (8).

The global finite element matrices in Eq. (23) may formally be written as

$$[K'] = \sum_e [K_e'] \quad [M'] = \sum_e [M_e'] \quad [R'] = \sum_e [R_e']$$

$$[C'] = \sum_e [C_e]^T [\beta_P][C_e] \quad \{f'\} = \sum_e \{f_e'\} \tag{24}$$

where Σ_e denotes the standard finite element assemblage operation [15].

8. Modeling of a point source

When the inhomogeneous Helmholtz equation in Eq. (1) is employed in the FE model, the source term involving the delta function, as the other terms of the differential equation, is satisfied only approximately over the finite elements in contact with the point source. Of course, the error is expected to decrease with mesh refinement. The present FE formulation uses the complex pressure p as the field variable. Hence, a kinematic/Dirichlet boundary condition in terms of p would be satisfied exactly at the finite element nodes. In light of this, it would be interesting to see whether the effect of the source could be modeled as a kinematic boundary condition. To facilitate this, the computational domain employed above (see Eq. (12)) excludes the source. This is achieved by matching each finite element node with the source, and excluding all the finite elements that are in contact with the source node. Then the free field pressure due to the source on the periphery of the excluded domain is imposed as a kinematic boundary condition in the finite element model. The discontinuity of the fields on the periphery of the region enclosing the source is our equivalent source. It may be argued that the pressure distribution on the excluded domain boundary is not the actual one, which would be known only after solving the FE equations. However, the following argument justifies the approach. It is known that for small volume sources, the pressure in the far field is not affected by the individual shape of a source, as long as the source strengths are equal. Thus, this justifies the use of a computational domain that excludes a small FE domain around a point source. In the present study, the size of the excluded domain has been kept at about a tenth of the wavelength. Comparison of the FE results with an analytical solution indicates that such a choice is satisfactory.

9. Solution of FE equations

The global FE equation in Eq. (23) may be written for brevity as

$$[A]\{\overline{p}'\} = \{f'\} \tag{25}$$

It may be noted that for an acoustic medium with real sound speed, the coefficient matrix $[A]$ above is complex even though $[K']$, $[M']$, and $[C']$ are real. This is because $\{f'\}$ is complex. Also, note that $[R']$ is complex due to the presence of α_m in Eq. (19d). For a lossy medium modeled with complex sound speed, $[M']$ is also complex. Although $[A]$ is non-self-adjoint, it is a complex symmetric matrix and hence the Gauss solver employed here to obtain the solution to Eq. (25) exploits the attendant computational advantage. Since such solvers for FE equations are coded as block solvers with compact storage scheme, large finite element models can be handled even with modest computer storage. Of course, such a solution strategy involves overhead in the form of read/write operations on secondary storage devices. This approach may be contrasted against those of Bayliss et al. [36] and Athanassoulis et al. [37] who have used iterative methods based on the conjugate-gradient technique. Solvers based on the conjugate-gradient method have been found much more efficient than Gauss solvers when the size of the matrix equation is very large, say, several tens of thousands of equations, and hence they hold promise for high frequency FE models.

Since the present FE model adopts a penalty function approach to impose the radiation boundary condition with multiple radiating modes, the choice of suitable penalty parameter α_{Pm} is important. This can be resolved through numerical experiments. The penalty parameter was obtained by prescribing a scale factor on the average value of the diagonals of the coefficient matrix $[A]$ in Eq. (25); i.e.,

$$\beta_{Pm} = \frac{\beta_s}{n'} \sum_{i=1}^{n'} |A_{ii}|, \tag{26}$$

where n' denotes the total number of FE equations/dof and β_s a user-specified penalty scale factor. Computations indicate that the results are stable over a wide range of β_s values. The results reported here have been obtained using $\beta_s = 100$.

10. Normal modes in an ocean waveguide with depth dependence

The sound speed in an ocean-acoustic waveguide is in general both depth- and range-dependent. Depth dependence is considered very important because it is responsible for many interesting phenomena in waveguide propagation. The two well-known methods that have been developed to study acoustic waves in depth-dependent waveguides are the fast-field technique and the normal-mode expansion [9, 1], the latter being the method that we have used. The normal-mode approach consists of first solving the depth eigenproblem for a given sound-speed profile to obtain the radial wavenumbers and the associated depth modes, which respectively are the eigenvalues and eigenfunctions. The depth eigenproblem could be solved exactly only for a few special profiles. In the finite element model for depth- and range-dependent waveguides [31-32], the eigensolution of the depth problem is required for imposition of the radiation condition at the truncation boundary. For such applications, it would be convenient to have the depth modes in a compact analytical form. We have explored this aspect with specific reference to shallow-water waveguides.

The depth eigenproblem can be cast in a variational form by suitably defining a functional. Then, the classical Rayleigh–Ritz method may be employed to find a variational approximation to the eigensolution of the depth problem in ocean-acoustic waveguides. The depth modes obtained would have a more compact analytical form than those derived using finite difference or finite element methods. The present work provides an RR model for the depth eigenproblem and demonstrates its utility for shallow-water waveguides.

11. Mathematical model

For the cylindrically symmetric waveguide having depth-dependent density ρ and sound speed c, the inhomogeneous pseudo Helmholtz equation governing the linear harmonic acoustic pressure field $p(r, z)$ in the waveguide is given in cylindrical coordinates (r,z) as [9, 1]

$$\frac{1}{r}\frac{\partial}{\partial r}\left(r\frac{\partial p}{\partial r}\right) + \rho(z)\frac{\partial}{\partial z}\left(\frac{1}{\rho(z)}\frac{\partial p}{\partial z}\right) + \frac{\omega^2}{c^2(z)}p = -\frac{\delta(r)\delta(z-z_s)}{2\pi r}, \tag{27}$$

where r denotes the range coordinate and z the depth coordinate as shown in Fig. 2, and the r.h.s. denotes a point source of unit amplitude located at $r = 0$ and $z = z_s$, with δ denoting the Dirac delta function. Eq. (27) can also be applied to problems with attenuation by introducing a complex sound speed.

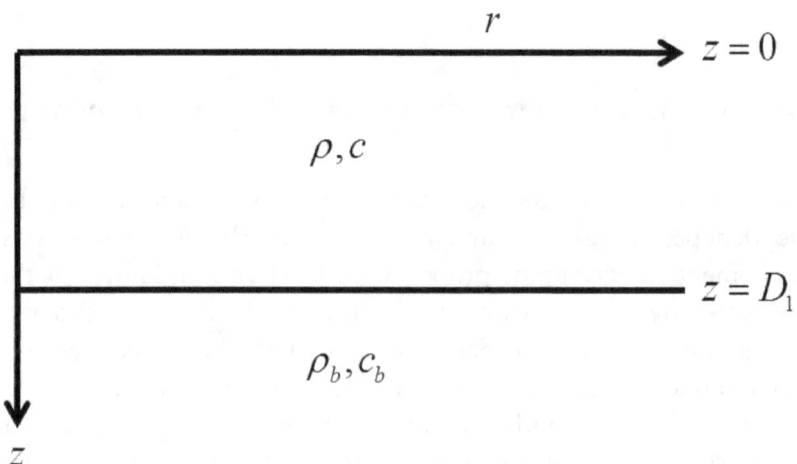

Figure 2. A two-layer cylindrically symmetric waveguide

A variable separable solution for the homogeneous form of Eq. (27) may be written as

$$p(r,z) = \bar{R}(r)Z(z) \tag{28}$$

Then, upon using Eq. (28) in the homogeneous form of Eq. (27), the following ordinary differential equations are obtained:

$$\frac{1}{r}\frac{d}{dr}\left(r\frac{d\bar{R}(r)}{dr}\right)+k_r^2\bar{R}(r)=0 \tag{29}$$

$$\rho(z)\frac{d}{dz}\left(\frac{1}{\rho(z)}\frac{dZ(z)}{dz}\right)+\left(\frac{\omega^2}{c^2(z)}-k_r^2\right)Z(z)=0 \tag{30}$$

where ω denotes the circular frequency and k_r^2, the separation constant, which turns out to be the square of the radial/horizontal wavenumber. Eq. (29) evidently pertains to the radial/horizontal modes $\bar{R}(r)$, and Eq. (30) pertains to the depth modes $Z(z)$. Choosing a pressure release boundary at the top $(z=0)$ and a mixed/Robin boundary condition at the seabed $(z=D_1)$, the boundary conditions of our problem are written as [1, 30, 33]

$$Z(0)=0 \tag{31}$$

$$Z(D_1)+\frac{g\left(k_r^2\right)}{\rho}\frac{dZ(D_1)}{dz}=0 \tag{32}$$

where

$$g(k_r^2)=\rho_b\,/\,\sqrt{\left(k_r^2-\frac{\omega^2}{c^2}\right)}$$

with ρ_b denoting the density of the acoustic fluid in the isovelocity half-space underlying the water column. Eq. (32) facilitates replacing the half-space in the Pekeris waveguide [38] by means of an impedance-type boundary condition. It may be noted that Eq. (30) together with the homogeneous boundary conditions in Eqs. (31) and (32) do not constitute a proper Sturm-Liouville problem because Eq. (32) depends on the unknown eigenvalue k_r^2. Porter and Reiss [30] employed a finite difference model to solve Eq. (30) together with the boundary conditions in Eqs. (31) and (32). As an alternative to the above formulation, the waves in the fluid half-space below the water column are also considered here [1]. The governing equation for the waves in this fluid half-space is given as,

$$\rho_b(z)\frac{d}{dz}\left(\frac{1}{\rho_b(z)}\frac{dZ_b(z)}{dz}\right)+\left(\frac{\omega^2}{c_b^2(z)}-k_r^2\right)Z_b(z)=0, \qquad D_1\le z\le\infty \tag{33}$$

where Z_b denotes the depth function in the fluid half-space having depth-dependent density ρ_b and sound speed c_b. The interface conditions at the seabed are given by the kinematic and dynamic conditions,

$$Z(D_1) = Z_b(D_1) \qquad\qquad\qquad (a)$$

$$\frac{1}{\rho}\frac{dZ(D_1)}{dz} = \frac{1}{\rho_b}\frac{dZ_b(D_1)}{dz} \qquad\qquad (b) \qquad\qquad (34)$$

In addition, the depth mode Z_b should remain bounded as $z \to \infty$. Our primary objective is to consider a variational formulation for Eqs. (30) and (33), together with appropriate boundary conditions, and obtain a RR approximation to the depth-dependent problem.

12. Variational formulation and Rayleigh–Ritz approximation

A variational formulation that leads to the boundary value problem in the last section is sought now. The operator being symmetric, there exists a functional, the variation of which leads to Eq. (30) and appropriate boundary conditions, and similarly for the half-space. Consider the functional $\Pi(Z)$ and $\Pi_b(Z_b)$ defined respectively in the water column and the half-space as

$$\Pi(Z) = \frac{1}{2}\int_0^{D_1}\left[\frac{1}{\rho(z)}\left(\frac{dZ}{dz}\right)^2 - \frac{\omega^2}{\rho(z)c^2(z)}Z^2 + \frac{1}{\rho(z)}k_r^2(z)Z^2\right]dz - \frac{1}{\rho(z)}Z_v Z\Big|_0^{D_1} \qquad (35)$$

$$\Pi_b(Z_b) = \frac{1}{2}\int_{D_1}^{D_2}\left[\frac{1}{\rho_b(z)}\left(\frac{dZ_b}{dz}\right)^2 - \frac{\omega^2}{\rho_b(z)c^2(z)}Z_b^2 + \frac{1}{\rho_b(z)}k_r^2 Z_b^2\right]dz$$

$$-\frac{1}{\rho_b(z)}Z_{bv}(D_1)Z_b(D_1) + \frac{1}{2}\beta\frac{Z_b^2(D_2)}{\rho_b}, \qquad D_2 \to \infty \qquad (36)$$

where suffix v denotes z-derivative. At the interface $z = D_1$ between the water column and the half-space, the conditions noted in Eq. (30) must be imposed. In view of Eq. (34b), this can be achieved by setting in Eq. (35)

$$\frac{1}{\rho}Z_v(D_1) = \frac{1}{\rho_b}\frac{dZ_b(D_1)}{dz}, \qquad\qquad (37)$$

and in Eq. (36),

$$\frac{1}{\rho_b}Z_{bv}(D_1) = \frac{1}{\rho}\frac{dZ(D_1)}{dz} \qquad\qquad (38)$$

In addition, Eq. (34a) should be imposed. Then, it can be easily shown that the variational condition $\delta \Pi = 0$ leads to Eq. (30) and the boundary conditions in Eq. (31) as well as the interface condition in Eq. (37), where δ denotes the first variation. Similarly, the variational condition $\delta \Pi_b = 0$ leads to Eq. (33), and the interface conditions in Eq. (34a) and Eq. (38). In addition, at $z = D_2$, we obtain the condition

$$\frac{dZ_b(D_2)}{dz} + \beta Z_b(D_2) = 0, \tag{39}$$

where (see Eq. (32)),

$$\beta = \sqrt{\left(k_r^2 - \frac{\omega^2}{c_b^2} \right)}$$

Note that there are three cases that can be analyzed using Eq. (36):

Case 1: D_2 is finite, $\beta \to 0$

This corresponds to the case when a depth-dependent seabed of finite thickness is terminated by a rigid boundary.

Case 2: β is finite, $D_2 \to \infty$

This corresponds to a depth-dependent seabed of infinite thickness.

Case 3: D_2 and β are finite

This is a three-layer problem, where the top layer is the water column, the second layer is a layer of seabed with depth varying density and speed, and the bottom layer represents the seabed of infinite extent with uniform sound speed and density.

We now seek an assumed mode solution with n terms to the above variational problem in the form

$$Z(z) \approx \sum_{j=1}^{n_1} \bar{\phi}_j \psi_j(z) \quad 0 \leq z \leq D_1 \quad \text{(a)}$$

$$Z_b(z) \approx \sum_{j=1}^{n_2} \bar{\phi}_{bj} \psi_{bj}(z) \quad D_1 \leq z \leq \infty \quad \text{(b)} \tag{40}$$

where $n = n_1 + n_2$, and ψ_j and ψ_{bj} denote the known mode function (coordinate function) satisfying the kinematic boundary condition in the water column and $\bar{\phi}_j$ an unknown constant, and their counterparts with suffix b correspond to those of the half-space. The two sets of mode functions above are such that they satisfy the relevant boundary conditions as well as the interface conditions in Eq. (34) and the conditions in Eq. (39). Such functions may readily be constructed by solving a two-layer depth problem, which is nothing but the Pekeris waveguide [38], with an appropriate choice of constant velocity and density in the water column and the seabed half-space. This approach has been adopted here. Then, it follows that the continuity of pressure field at the interface $z = D_1$ implies that the assumed mode expansion in Eq. (40) reduces as,

$$Z(z) \approx \sum_{j=1}^{n} \bar{\phi}_j \psi_j(z) \quad 0 \le z \le D_2 \tag{41}$$

where we have combined the depth modes of a two-layer isovelocity waveguide as one combined set with redefined coefficients $\bar{\phi}$. Then, using Eq. (40) in the functionals in Eqs. (35) and (36), and combining them, an algebraic approximation for the functionals is obtained as

$$\bar{\Pi} = \Pi + \Pi_b = \frac{1}{2}\{\bar{f}\}^T \left[K^I\right]\{\bar{f}\} - \frac{1}{2}\omega^2 \{\bar{f}\}^T \left[K^{II}\right]\{\bar{f}\} + \frac{1}{2}k_r^2 \{\bar{f}\}^T \left[K^{III}\right]\{\bar{f}\} \tag{42}$$

where

$$K_{ij}^{I} = \int_0^{D_1} \frac{1}{\rho(z)} \frac{d\psi_i}{dz} \frac{d\psi_j}{dz} dz + \int_{D_1}^{D_2} \frac{1}{\rho_b(z)} \frac{d\psi_{bi}}{dz} \frac{d\psi_{bj}}{dz} dz$$

$$K_{ij}^{II} = \int_0^{D_1} \frac{1}{\rho(z)c^2(z)} \psi_i \psi_j dz + \int_{D_1}^{D_2} \frac{1}{\rho_b(z)c_b^2(z)} \psi_{bi} \psi_{bj} dz \tag{43}$$

$$K_{ij}^{III} = \int_0^{D_1} \frac{1}{\rho(z)} \psi_i \psi_j dz + \int_{D_1}^{D_2} \frac{1}{\rho_b(z)} \psi_{bi} \psi_{bj} dz$$

It has been assumed in the above that the contribution due to the term in Eq. (36) is negligible as $D_2 \to \infty$. Further, it may be noted that since the boundary and interface conditions are satisfied by the trial functions chosen above, when the functionals in Eqs. (35) and (36) are combined to obtain Eq. (42), the boundary and interface terms add up to become trivial and hence do not contribute to the discrete approximation in Eq. (42).

The variational condition is now replaced by the condition

$$\frac{\partial \bar{\Pi}}{\partial \phi_j} = 0, \quad j = 1, 2, \dots, n \tag{44}$$

Eq. (44) yields a symmetric algebraic eigenproblem given by

$$\left(\omega^2\left[K^{II}\right]-\left[K^{I}\right]\right)\{\bar{\phi}\}=k_r^2\left[K^{III}\right]\{\bar{\phi}\}$$

(45)

The eigensolution of Eq. (45) may be denoted as

$$\left(k_{rj}^2,\left\{\phi^{(j)}\right\}\right),\qquad j=1,2,..n$$

(46)

It may be noted here that the eigenproblem in Eq. (45) remains linear unlike the Porter and Reiss model that is based on Eqs. (30)–(32). Having obtained the eigenvalues k_{rj}^2 and the eigenvectors $\{\bar{\phi}^{(j)}\}$, the eigenfunctions /depth modes may be written, using Eq. (40), as

$$Z_j(z)=\left\{\bar{\phi}^{(j)}\right\}^T\{\psi(z)\}$$

(47)

where

$$\{\psi(z)\}^T=\langle\psi_1(z),\psi_2(z),\cdots,\psi_n(z)\rangle$$

(48)

Eq. (47) provides a compact *semi-analytical form* for the depth modes that are convenient to employ in FE models such as those in Fix and Marin [31] and Vendhan et al. [32] for approximating the radiation condition at the truncation boundary. The depth modes obtained can of course be used to set up the normal-mode solution to the forced Helmholtz equation in Eq. (27). Note that for a Pekeris waveguide, the normal-mode solution based on the discrete spectrum has to be augmented with the continuous spectrum contribution [1]. Since the eigenvectors in Eq. (45) are $[K^{III}]$ -orthogonal, it can easily be shown that the eigenfunctions in Eq. (47) satisfy the following orthogonality condition:

$$\int_0^{D_1}\frac{1}{\rho(z)}Z_iZ_j dz+\int_{D_1}^{D_2}\frac{1}{\rho_b(z)}Z_iZ_j dz=0\,,\qquad i\neq j$$

(49)

The orthonormal depth functions are obtained as

$$\bar{Z}_j(z)=Z_j/\sqrt{\left\{\bar{\phi}^{(j)}\right\}^T\left[K^{III}\right]\left\{\bar{\phi}^{(j)}\right\}}$$

(50)

In terms of finite element terminology, the RR model for each layer may be looked upon as a super-element with C_1 continuity at the inter-element boundary and the operation leading to Eq. (42) is equivalent to element-assemblage operation.

13. Numerical analysis and discussion

In our Rayleigh–Ritz model, the first task is to compute the symmetric matrices $[K^I]$, $[K^{II}]$, and $[K^{III}]$ in Eq. (43). The next task is to find the eigensolution to Eq. (45). For problems with no attenuation, the real eigenvalues have been obtained employing the bisection method. For problems with attenuation, approximations to the complex roots have been obtained using a search procedure [39] and the eigenvalues refined by employing Newton–Raphson iteration. In all cases, the eigenvectors are obtained using inverse iteration.

To validate our algorithm, we applied the Rayleigh–Ritz model first to single-layer isovelocity waveguide examples without attenuation for which exact solutions are available. Different sound-speed profiles have been chosen to evaluate the accuracy of the RR model. Attenuation in the fluid half-space has also been considered. Different sets of RR approximations have been obtained by varying the number of assumed modes n in Eq. (41). The results for $n = 2n_p$, where n_p denotes the number of propagating modes turned out to be of good accuracy.

One should note the following remarks in connection with the performance of the RR model for the depth eigenproblem:

a. The mode shapes of an isovelocity waveguide have been chosen as trial functions, which satisfy appropriate interface conditions and the condition at the free surface. This renders the RR matrix highly diagonally dominant, which also greatly aids in numerical evaluation of the eigensolution.

b. For ocean waveguides, the depth variation of the sound speed is normally only a small percentage of the unperturbed value.

c. When the variation in sound speed is large, the above procedure may not give good results. One has to resort to high-order solutions. Even then, one can expect accurate eigenvalues, but not eigenvectors. This is because the convergence rate for the eigenvectors is slower than that for the eigenvalues.

14. Numerical examples

We considered several examples to illustrate the versatility of our FEM approach in remote sensing problems. In all our examples, we employed a Dirichlet boundary condition on the air–sea interface, a Neumann boundary condition on the ocean bottom boundary, and a unit point source at a depth of 36 m below the air–water interface. Both depth-dependent and uniform sound-speed water columns are considered.

14.1. Isovelocity case

The finite element method for the solution of inhomogeneous ocean-acoustic waveguide problems is validated first with analytical results for isovelocity waveguides. A cylindrically symmetric plane parallel waveguide of depth 100 m with a point source is shown in Fig. 3. The finite element model consists of a uniform grid of isoparametric quadrilateral elements, with the element length being about a tenth of the source signal wavelength. As discussed previously, a domain of two elements has been excluded to remove the source from the truncated domain (Fig. 1). The FE mesh consists of 1000 elements in range and 60 elements in depth. The computed acoustic pressure along the range at the depth of the source is compared in Fig. 3 with the normal-mode solution with 50 modes, of which only the leading few modes are propagating. In all cases, the FEM results compared well with analytical results. The mesh is chosen appropriately so that the modal error is less than 5%.

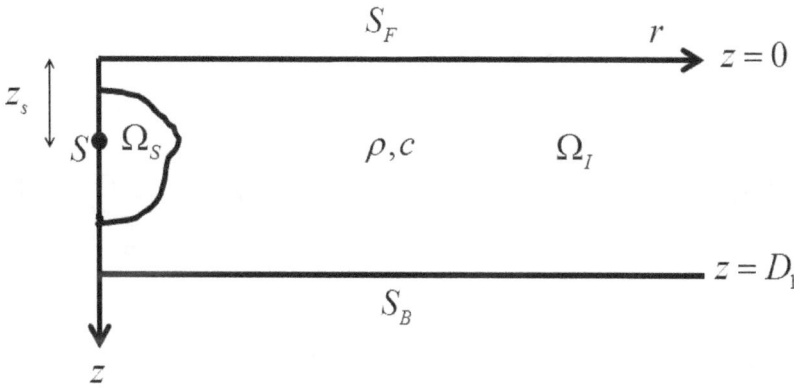

Figure 3. Idealized ocean waveguide

14.2. Rectangular hump

Sea mounts are often encountered in under-water ocean problems. In order to understand their impact on wave propagation characteristics in shallow-water environment, we considered a rigid rectangular hump of width 40 m and height 20 m on ocean bottom as shown in Fig. 4(a). The contour map of transmission loss (TL) of a 60-Hz point source located on the z-axis at a depth of 36 m from the water surface is shown in Fig. 4. Panel (a) shows the TL in the presence of a rectangular hump on the seabed. Panel (b) shows the TL of the water column without the rectangular hump. Notice that the rectangular hump has a distinct signature in TL pattern especially on the right of the hump.

It is instructive to take a look at the acoustic power distribution in the modes. Figure 5 shows the modal power spectrum of the shallow-water column with the rectangular hump in panel (a) and without the rectangular hump in panel (b). Notice that there is a substantial redistribution of power among the modes due to the presence of the rectangular hump.

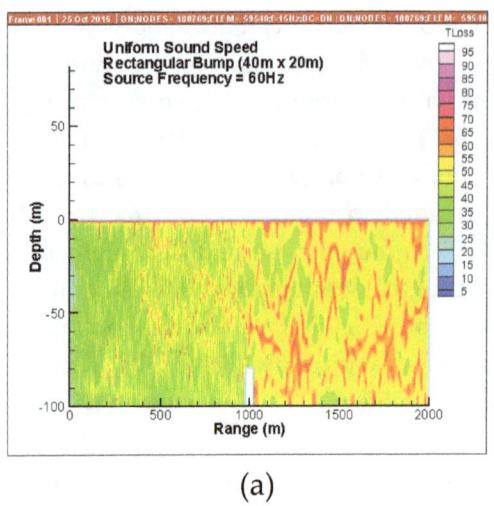 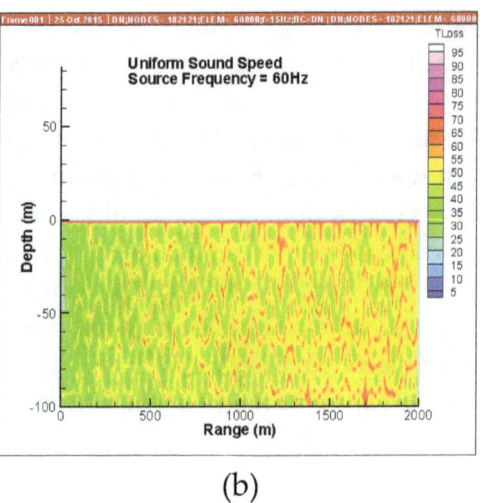

Figure 4. Transmission loss of a shallow-water column with a (a) rectangular hump on the seabed and (b) flat-bottom surface

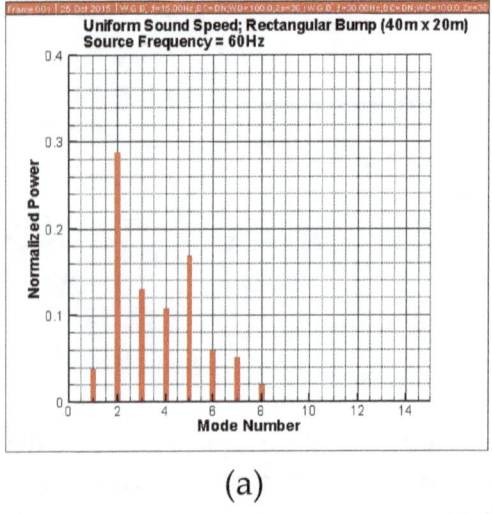 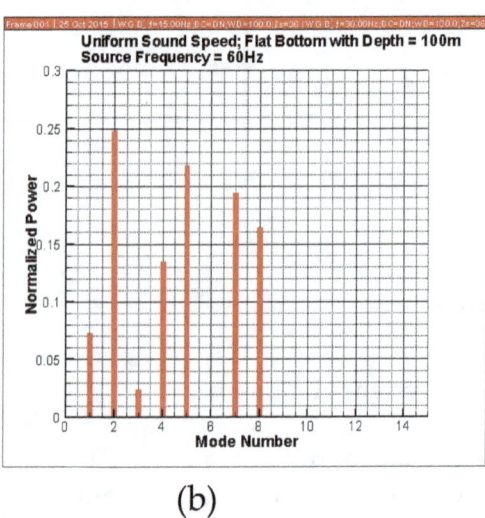

Figure 5. Spectra of modal efficiencies for (a) seabed with a rectangular bump and (b) a flat seabed

14.3. Down-sloping bottom

Shallow-water conditions are encountered in the near-coast context where the ocean bottom has a sloping geometry. There are two situations to consider, up-slope and down-slope, depending on the location of the source with respect to the slope. First we consider the down-sloping case where the ocean-bottom slopes down from 100 m to 230 m over a distance of 600 m. The details of the geometry are shown in Fig. 6.

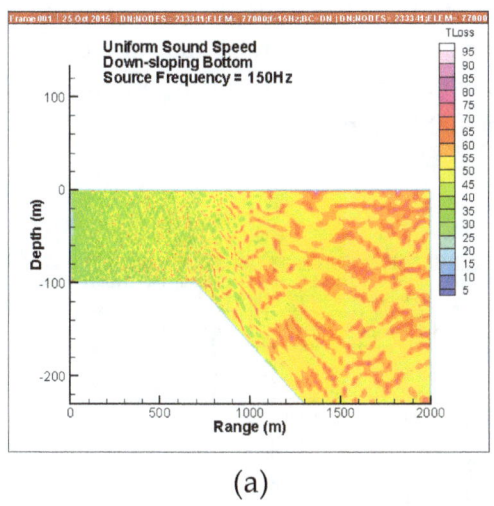
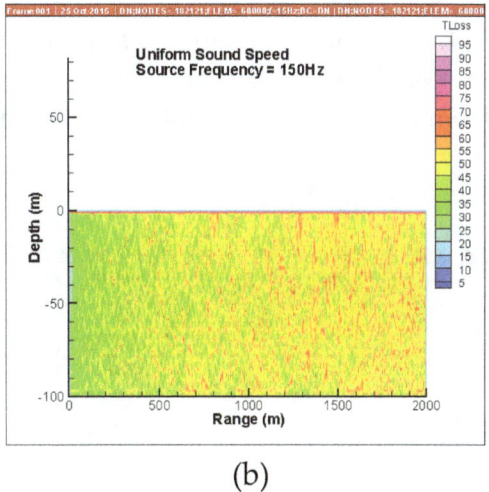

(a)   (b)

Figure 6. Transmission loss of a shallow-water column with (a) down-sloping bottom and (b) flat bottom of depth 100 m

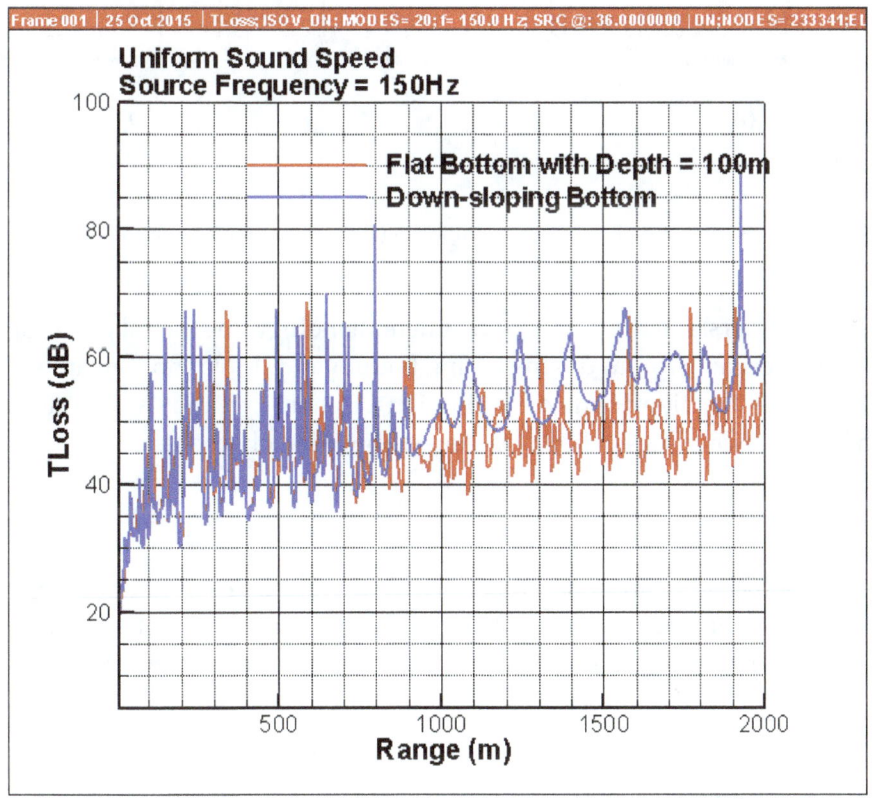

Figure 7. Comparison of TL at 36 m depth of an ocean with down-sloping bottom and a flat bottom

Panel (a) shows the TL with the down-sloping bottom. Panel (b) shows the TL for a water column with the flat bottom at depth 100 m. Both results are for the source frequency of 150 Hz. Notice the distinct spatial power distribution manifested by the sloping bottom. To facilitate a better comparison, we have shown in Fig. 7 the TL at 36 m depth corresponding to the flat and sloping bottoms. Notice that the TLs for the two cases are similar in the region between the source and the middle of the slope. Beyond that, the TL corresponding to the sloping bottom is significantly larger than that of the flat bottom.

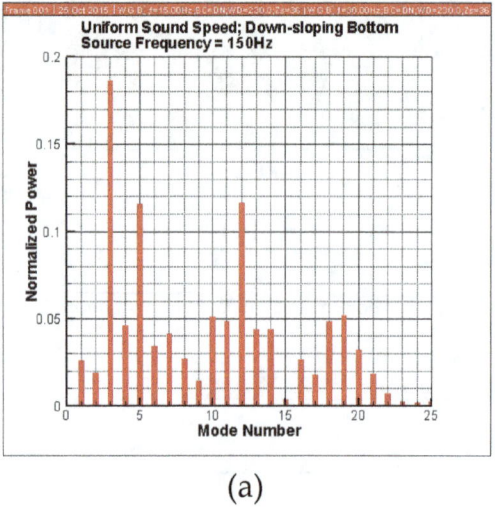
(a)

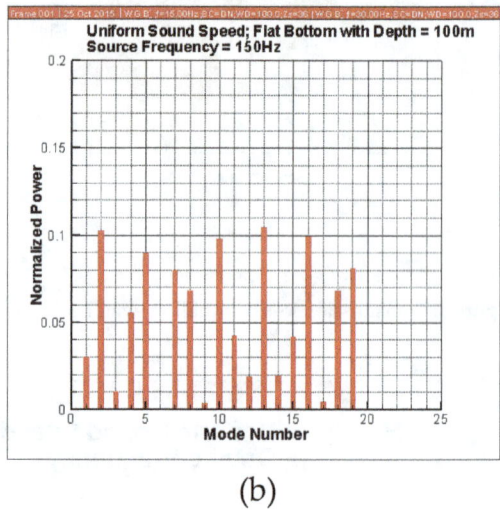
(b)

Figure 8. Modal power spectrum of shallow-water column with (a) down-sloping bottom and (b) flat bottom of depth 100 m

In order to better understand the propagation phenomenology, the modal power spectrum for the shallow-water ocean with (a) down-sloping bottom and (b) flat bottom are shown in Fig. 8. Notice that there is a significant redistribution of energy in the case of sloping bottom although the total power flows in both cases are approximately the same.

14.4. Up-sloping bottom

Next we consider the problem of sloping bottom where the ocean bottom slopes up (with respect to location of the source) from 230 m to 100 m over a distance of 600 m. The details of the geometry are shown in Fig. 9. The acoustic source is located at 36 m below the water surface on the left.

Panel (a) shows TL for the case of 105Hz and panel (b) shows the case of 150 Hz. We notice that at 105 Hz there is a substantial reduction in power flow. However, at 150 Hz the power flow is as good as that of a flat-bottom waveguide. The mode spectral distribution in Fig. 10 shows the details of how the power flows in the two cases. We notice that for the up-slope case, power flow can be good at certain frequencies and not good at others, depending on the impedance matching conditions. In contrast, for the case of down slope the power flow is good for all the frequencies that we studied.

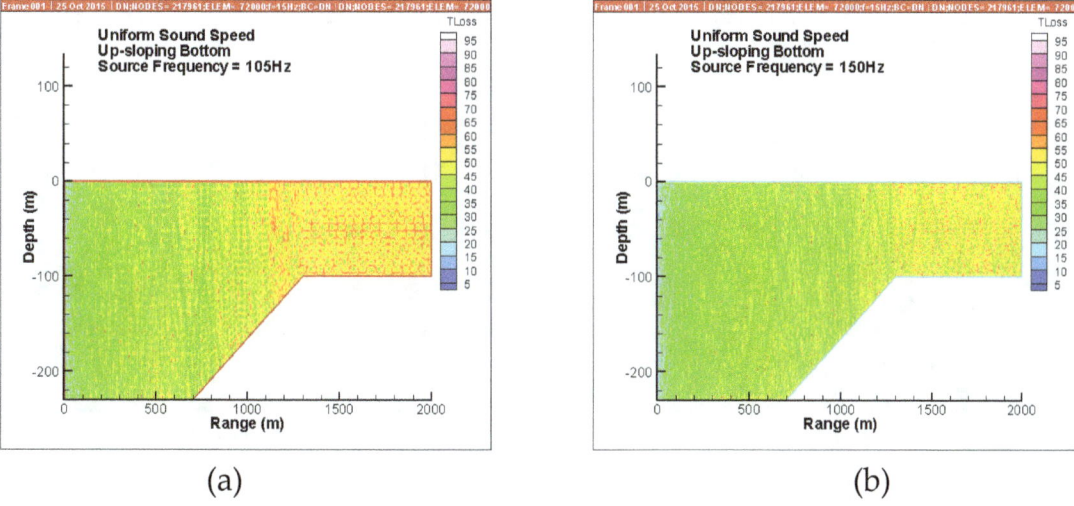

(a)   (b)

Figure 9. Transmission loss of shallow-water column with up-slope bottom at 105 Hz and 150 Hz

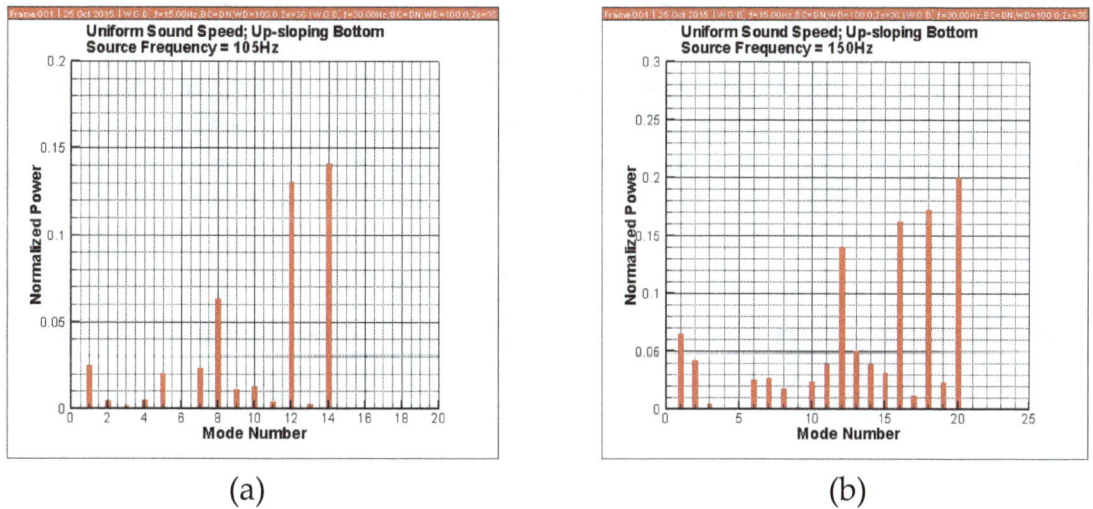

(a)   (b)

Figure 10. Modal spectrum of shallow-water column with up-sloping bottom (a) 105 Hz (b) 150 Hz

14.5. Object in the water column

Characterizing the signatures of objects in the ocean is an important remote sensing problem. We consider a cylindrical rigid object of radius 20 m in the middle of a water column as shown in Fig. 11. Panel (a) shows the TL for the source frequency at 135 Hz. Panel (b) shows the TL for the source frequency at 150 Hz. We notice that the power flow can be substantially influenced by the object, depending on the frequency of operation. This is because of the interference phenomena involving the object and the boundaries of the waveguide.

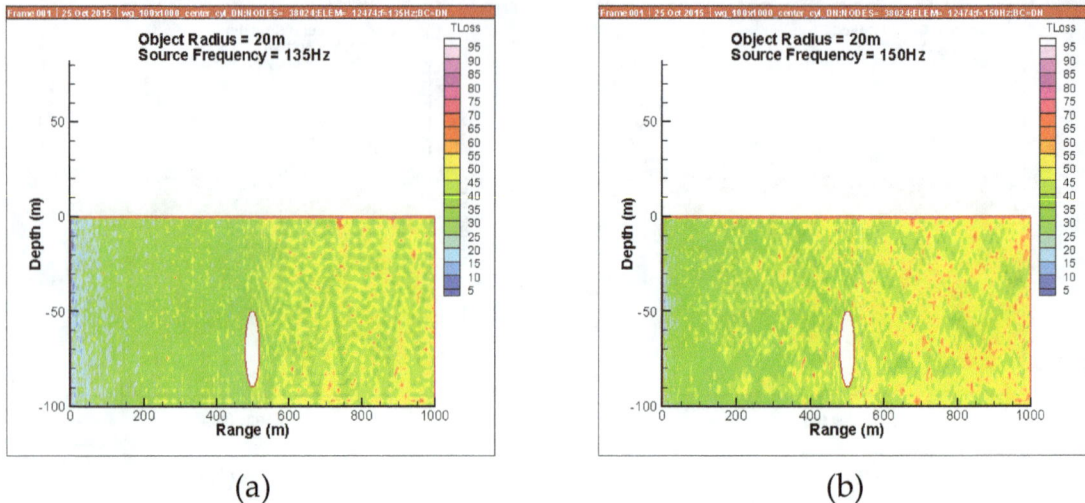

(a)   (b)

Figure 11. TL of a shallow-water column with an object (a) 135 Hz (b) 150 Hz

14.6. Shallow-water column with rippled top surface

Ripples on the water surface can be generated by gravity and wind conditions. Such surface undulations can considerably influence the wave propagation in the shallow-water wave-guide. To illustrate this phenomenon, we have taken a periodic structure on the air–water interface as shown in Fig. 12. The top surface has a sinusoidal undulation of amplitude 5 m and period 50 m.

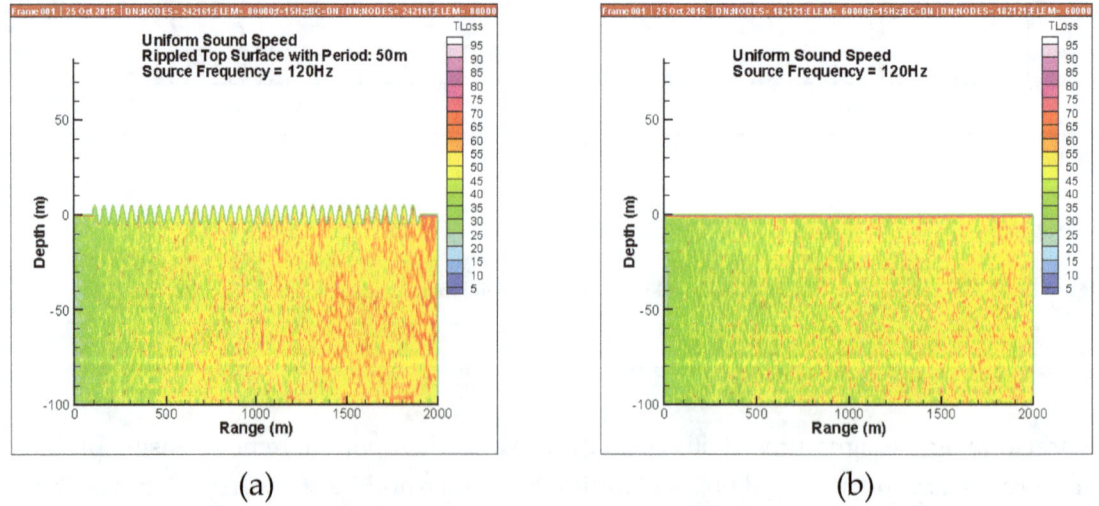

(a)   (b)

Figure 12. TL in a shallow-water column with (a) wind-generated rippled air–water interface, and (b) flat water surface

Our FEM results show that the surface ripples causes substantial transmission loss compared to that of the flat water surface for the case when the source frequency is 120 Hz. However, this pattern is quite sensitive to source frequency. For some frequencies, the TL may be large and for others, TL can be low. Dimensions of the waveguide and ripple geometry in terms of the source signal wavelength are key factors influencing the physics.

14.7. Shallow-water column with depth-dependent sound speed

In all the examples considered thus far, we have assumed that the water column has uniform sound speed. This is rarely true in practice even for the shallow-water ocean. The normal modes for the depth-dependent waveguide are required to impose the radiation boundary condition in our finite element procedure. The Rayleigh–Ritz approximation is used for obtaining normal modes for this problem. The sound-speed profile taken for this study is shown in Fig. 13.

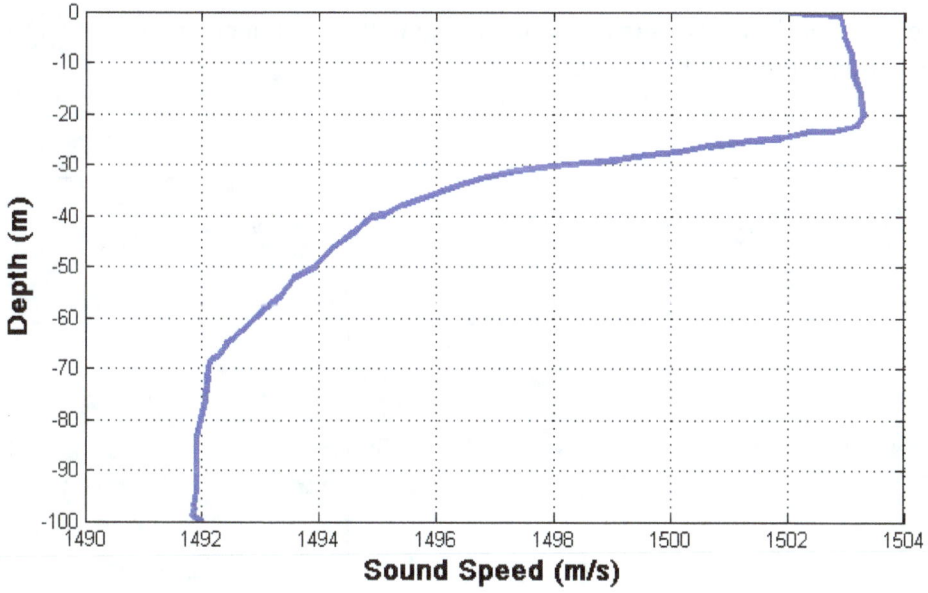

Figure 13. Sound-speed profile (S1) used for our study

The TL for our geometry with the sound-speed profile given in Fig. 13 is shown in Fig. 14. The result for source frequency of 150 Hz is shown in Panel (a) and that corresponding to isovelocity is shown in Panel (b). Although the sound-speed variation is very small, we notice the impact of depth dependence of sound speed on TL is substantial. However, at lower frequencies, this kind of sound-speed variation does not influence the TL much.

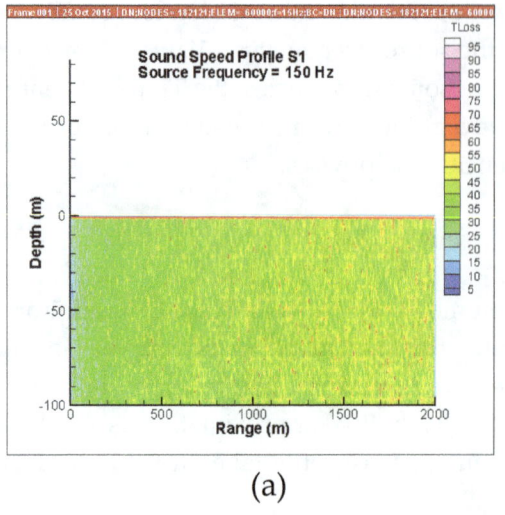
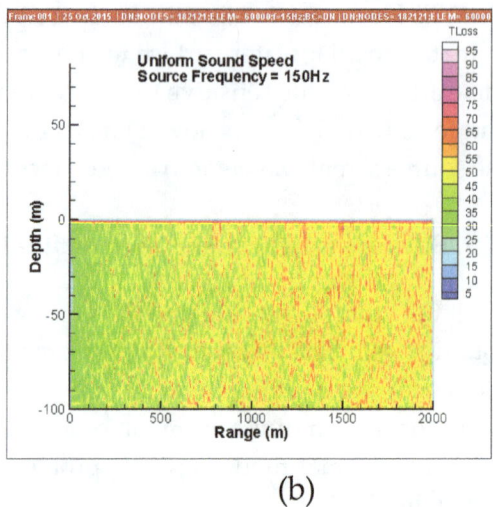

(a)   (b)

Figure 14. TL for the shallow-water ocean with (a) depth-dependent sound speed, and (b) uniform sound speed

14.8. Shallow-water column with depth-dependent sound speed and a rectangular bump on seabed

Next, we consider the case of shallow-water ocean with depth-dependent sound speed and a rigid rectangular hump on the seabed.

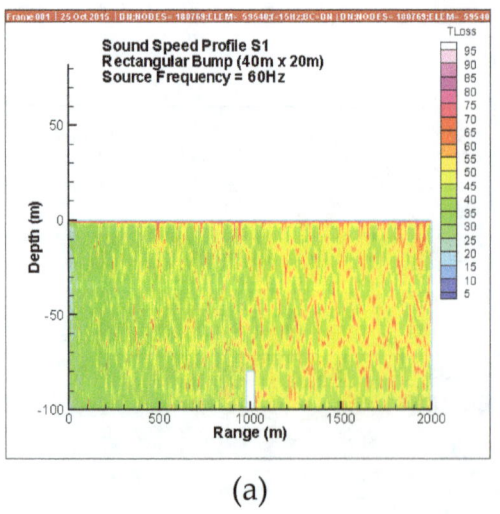
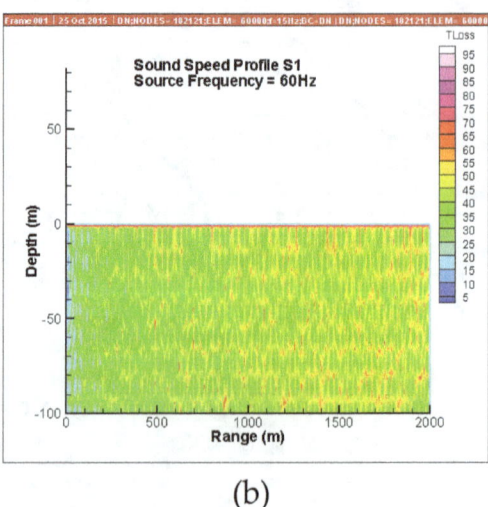

(a)   (b)

Figure 15. TL for the shallow-water ocean with (a) depth-dependent sound speed and a rectangular bump on the seabed, and (b) depth-dependent sound speed and flat bottom

We observe that, for the case when the source frequency is 60Hz, the presence of the small rectangular hump on the seabed has a significant effect on the transmission loss of a depth-dependent ocean.

14.9. Shallow-water column with depth-dependent sound speed and rippled top surface

Finally, we consider the case of shallow-water ocean with depth-dependent sound speed (Fig. 13) and a rippled air–water surface. The results are shown in Fig. 16.

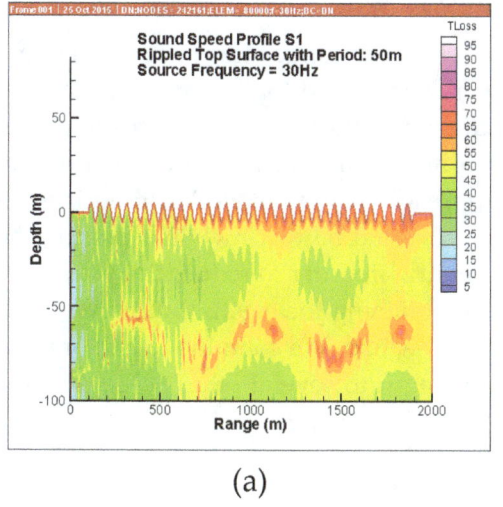
(a)

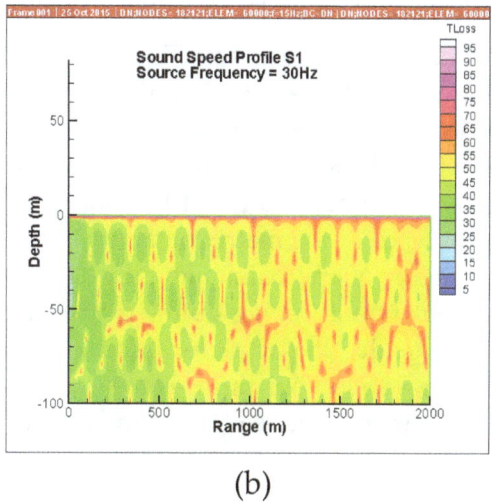
(b)

Figure 16. TL for the shallow-water ocean with (a) depth-dependent sound speed and a rippled air–water interface, and (b) depth-dependent sound speed

Note that for source frequency of 30Hz, the presence of ripples has reduced the transmission loss in most regions. This is in contrast to the last case (Fig. 15) where there is a rectangular bump on the seabed. However, these characteristics are due to interference phenomenon and hence have strong frequency dependence. The important point is that small features such as ripples can have a significant impact on the underwater propagation characteristics.

15. Conclusion

A finite element approach has been presented for remote sensing in shallow-water ocean environment. The three principal elements of remote sensing are: (a) signal propagation and reception, (b) data analysis, and (c) inversion or retrieval. This chapter exclusively deals with part (a) of the trilogy of remote sensing. Although several approaches have been developed for wave propagation studies in underwater ocean, they all have limitations when encountered with complex geometries and environments as in shallow-water ocean. An FE approach is both accurate and feasible for such applications. In order to minimize the problem size, a Bayliss-type damper was imposed to truncate the solution domain. Since several propagating modes can exist in the ocean waveguide, a penalty function approach was used to impose the radiation boundary condition in the variational finite element formulation of the problem. This penalty function approach was found to be robust over a wide range of penalty scale factors.

For the shallow-water ocean waveguide with depth-dependent sound-speed problem, the eigensolution was obtained using a Rayleigh–Ritz approximation. The trial functions are derived from an isovelocity problem that has exact solution. It is important to note that such trial functions automatically satisfy even the dynamic interface condition at the seabed, thus contributing to the accuracy of the numerical model. The proposed model is accurate and provides a compact semi-analytical form for the depth modes.

We thus have an accurate FE model for the remote sensing in range- and depth-dependent ocean-acoustic waveguides. Numerous examples were considered to illustrate the accuracy and versatility of this model. Admittedly, the computational effort in setting up the matrix in the proposed RR model using numerical quadrature is high compared to setting up the finite-difference-based matrix in the Porter and Reiss approach. However, noting the diagonal dominance of the matrix obtained in the RR model, it would be worthwhile exploring the possibility of approximating it by a narrow banded matrix in order to reduce the volume of computation in setting up the matrix and possibly in obtaining the eigensolution. We have also extended this work for the case of irregular elastic seabed. We continue to explore and further develop our finite element approach by applying it to several other ocean-acoustic problems encountered in the remote sensing of ocean environment.

16. Appendix A: Derivation of multimode radiation damping matrix

Consider the functional in Eq. (12). The contribution, $I_R(p_e)$, from the radiation boundary of a finite element is represented by the second integral in that equation; i.e.,

$$I_R(p_e) = \frac{1}{2} \sum_{m=1}^{M} \int_{S_{Re}} \frac{1}{\rho} \alpha_m p_m^2 \, r dz \tag{51}$$

where M denotes the number of propagating modes, α_m the damper coefficient associated with the m-th mode [see Eq. (5)], $p_m(z)$ the pressure associated with the m-th normal mode, and S_{Re} the element surface on the radiation boundary (see Fig. 1).

The modal pressure on the radiation boundary is given by Eq. (8):

$$p_m(z) = a_m f_m(z) \qquad m = 1, 2, \dots, M, \tag{52}$$

where $f_m(z)$ denotes the normal-mode function and a_m the modal coefficient. Using the finite element representation, the modal pressure on the radiation boundary may be written as

$$p_m(z) = \sum_e [N]^T \{\bar{p}_{em}\}, \tag{53}$$

where $[N]$ denotes the shape functions and $\{\bar{p}_{em}\}$ the nodal pressure vector on an element edge on the radiation boundary due to the m-th mode. The summation symbol is used to indicate that Eq. (53) is a piecewise polynomial representation over the entire depth of the waveguide. Using Eqs. (52) and (53), Eq. (51) may be written in a discrete form for a finite element as [also see Eq. (15)]

$$I_R(p_e) = \frac{1}{2}\sum_{m=1}^{M}\{\bar{p}_{em}\}^T\{\bar{R}_{em}\}\{p_{em}\},\tag{54}$$

where

$$\left[\bar{R}_{em}\right] = \alpha_m \int_{S_{Re}} \tfrac{1}{\rho}\left[N\right]^T\left[N\right]dS.\tag{55}$$

In view of Eq. (52), the vector of modal pressure at the nodes of an element in Eq. (53) may be written as

$$\{\bar{p}_{em}\} = a_m\left\langle f_{zm}(z_1), f_{zm}(z_2),\ldots, f_{zm}(z_n)\right\rangle^T = a_m\{f_{zm}\},\tag{56}$$

where $f_{zm}(z_j)$ denotes the j-th nodal value of the m-th eigenmode on a finite element in contact with the radiation boundary. Now, using Eqs. (55) and (56), the functional in Eq. (54) may be written as

$$I_R = \frac{1}{2}\sum_{m=1}^{M}a_m^2 R_{em},\tag{57}$$

where

$$R_{em} = \{f_{zm}\}^T\left[\bar{R}_{em}\right]\{f_{zm}\}.\tag{58}$$

The foregoing steps form the basis for Eq. (19c).

17. Appendix B: Normal-mode functions for isovelocity waveguides

The Rayleigh–Ritz model presented for the depth eigenproblems employs the analytical depth modes for an isovelocity waveguide as the trial functions. The details of the various isovelocity waveguide examples encountered in the ocean context are presented here. It should be kept in mind that in our problem, the acoustic source and reception points are both in the water column. Therefore, the wave functions given here have been chosen particularly for this application.

For a single-layer waveguide of depth D with Dirichlet boundary condition on top and Neumann boundary condition on the bottom surface, the trial functions are given by

$$\psi_j = a_j \sin(k_{zj} z) \tag{59}$$

where

$$k_{zj} = \frac{(j - 0.5)\pi}{D}, j = 1, 2, \ldots \tag{60}$$

where a_j is chosen to normalize the mode functions.

For a two-layer waveguide shown in Fig. 17, the trial functions are given by

$$
\begin{aligned}
\psi_j(z) &= a_j \sin k_{zj} z & 0 \leq z \leq D_1 \\
&= a_j \frac{\sin k_{zj} D_1}{\cos k_{bzj}(D_2 - D_1)} \cos k_{bzj}(D_2 - z) & D_1 \leq z \leq D_2
\end{aligned} \tag{61}
$$

where $k_{zj}^2 = \omega^2/c^2 - k_{rj}^2$, $k_{bzj}^2 = \omega^2/c_b^2 - k_{rj}^2$ and k_{rj} is the solution of the transcendental equation

$$\frac{k_{zj}\rho_b}{k_{bzj}\rho} = \tan k_{zj} D_1 \tan k_{bzj}(D_2 - D_1) \tag{62}$$

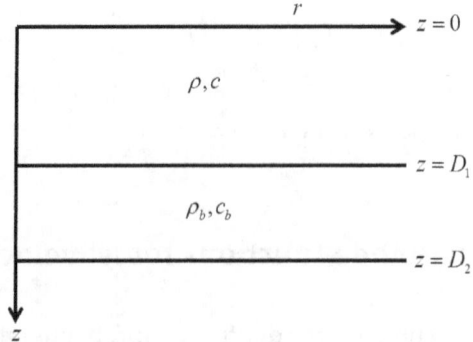

Figure 17. A bounded two-layer waveguide

In the case of a Pekeris waveguide (for which $D_2 \to \infty$ in Fig. 17), the trial functions are given by

$$
\begin{aligned}
\psi_j(z) &= a_j \sin k_{zj} z & 0 \le z \le D_1 \\
&= a_j \sin(k_{bzj} D_1) e^{-i k_{bzj}(D_1 - z)} & D_1 \le z \le \infty
\end{aligned}
\tag{63}
$$

where $k_{zj} = \sqrt{\omega^2/c^2 - k_{rj}^2}$, $k_{bzj} = i\sqrt{k_{rj}^2 - \omega^2/c_b^2}$ and k_{rj} is the solution of the transcendental equation

$$
\tan k_{zj} D_1 = -\frac{i \rho_b k_{zj}}{\rho k_{bzj}}
\tag{64}
$$

Note that discrete guided modes in the water column exist only for the case when $k_{rj}^2 > \omega^2/c_b^2$.

There are two cases to consider:

Case 1: $k < k_b$

This applies to the situation when the sound-speed velocity in the seabed is smaller than that in the water column. In this case, there are no guided modes. The entire spectrum is continuous and does not have contribution to sound transmission in the water column at long distances.

Case 2: $k > k_b$

This applies to the situation when the sound speed in the seabed is larger than that in the water column. Here the spectrum consists of (a) discrete guided modes, (b) continuous radiation modes, and (c) surface modes. Among the three, it is the discrete guided modes that carry the sound signal over long distances in the water column.

Since our interest is in long-range sound transmission in the water column, we have restricted attention to discrete guided modes as shown above.

One should observe that our two-layer waveguide problem does not share the above mentioned behavior. Note that the two-layer waveguide is terminated at the bottom by a rigid boundary. Therefore, the underlying physical processes are different.

Case 1: $k < k_b$

Here the entire spectrum in the waveguide consists of discrete guided modes.

Case 2: $k > k_b$

In this case, the spectrum consists of discrete guided modes and surface modes. However, for long range propagation, the modes of significance are the discrete guided modes.

Acknowledgements

S. Mudaliar thanks the AFOSR for support. C.P. Vendhan thanks AOARD for support during this work period. The authors thank A.D. Chowdhury for his help in eigenvalue computations for depth-dependent water column.

Author details

Saba Mudaliar[1*], C.P. Vendhan[2] and C. Prabavathi[3]

*Address all correspondence to: saba.mudaliar@us.af.mil

1 Sensors Directorate, Air Force Research Laboratory, WPAFB Ohio, USA

2 Indian Institute of Technology Madras, Chennai, India

3 Independent researcher, Dayton, Ohio, USA

References

[1] Jensen, F.B., W.A. Kuperman, M.B. Porter, and H. Schmidt (2011) *Computational Ocean Acoustics*, 2nd ed., Springer, New York.

[2] Lurton, X. (2010) *An Introduction to Underwater Acoustics*, 2nd ed., Springer, Berlin.

[3] Medwin, H. (2005) *Sounds in the Sea*, Cambridge University Press, Cambridge.

[4] Urick, R.J. (1983) *Principles of Underwater Sound*, 3rd ed., McGraw-Hill, New York.

[5] Brekhovskikh, L. and Y. Lysanov (1982) *Fundamentals of Ocean Acoustics*, Springer-Verlag, Berlin.

[6] Buckingham, M.J. (1992) Ocean-acoustic propagation models, *J. Acoustique*, Vol. 5, pp. 223–287.

[7] Harrison, C.H. (1989) Ocean acoustic propagation models, *App. Acous.*, Vol. 27, pp. 163–201.

[8] Westwood, E.K. (1990) Ray model solutions to the bench, *J. Acous. Soc. Am.*, Vol. 87, pp. 1539–1545.

[9] Frisk, G.V. (1994) *Ocean and Seabed Acoustics: A Theory of Wave Propagation*, Prentice Hall, New Jersey.

[10] Evans, R.B. (1983) A coupled mode solution for acoustic propagation in a waveguide with stepwise depth variations of a penetrable bottom, *J. Acous. Soc. Am.*, Vol. 74, pp. 188–195.

[11] Tappert, F.D (1977) The parabolic approximation method, wave propagation and underwater acoustics, in *Wave Propagation and Underwater Acoustics*, J.B. Keller and J.S. Papadakis, eds., Springer-Verlag, Berlin, pp. 224–287.

[12] DiNapoli, F.R. and R.L. Deavenport (1980) Theoretical and numerical Green's function field solution in a plane multilayered medium, *J. Acous. Soc. Am.*, Vol. 67, pp. 92–105.

[13] Zienkiewicz, O.C. and R.L. Taylor (1989) *The Finite Element Method, Vol. 1- Basic Formulation and Linear Problems*, McGraw-Hill, New York.

[14] Bathe, K.J. (1982) *Finite Element Procedures in Engineering Analysis*, Prentice Hall, Inc., Englewood Cliffs, New Jersey.

[15] Cook R.D., D.S. Malkus, M.E. Plesha, and R.J. Witt (2002) *Concepts and Applications of Finite Element Analysis*, 4th ed., John Wiley, New York.

[16] Schmidt, H. and F.B. Jensen (1985) A full wave solution for propagation in multilayered viscoelastic media with application to Gaussian beam reflection and fluid-solid interfaces, *J. Acous. Soc. Am.*, Vol. 77, pp. 813–25.

[17] Thompson, D.J. and N.R. Chapman (1983) A wide-angle split-step algorithm for the parabolic equation, *J. Acous. Soc. Am.*, Vol. 74, pp. 1848–1854.

[18] Lee, D., G. Botseas, and J.S. Papadakis (1981) Finite difference solution to the parabolic wave equation, *J. Acous. Soc. Am.*, Vol. 70, pp. 795–800.

[19] Huang, D. (1988) Finite element solution to the parabolic wave equation, *J. Acous. Soc. Am.*, Vol. 84, pp. 1405–1413.

[20] Evans, R.B. and K.E. Gilbert (1985) Acoustic propagation in a reflecting ocean waveguide with an irregular interface, *Comp. Math. Appls.* Vol. 11, pp. 795–805.

[21] Dougalis, V.A., N.A. Kampanis, and M.I. Taroudakis (1998) Comparison of finite element and coupled mode solutions of the Helmholtz equation in underwater acoustics, *Proc. 4th European Conference in Underwater Acoustics*, eds. A. Alippi and G.B. Cannelli, CNR-IDAC, Rome, Vol. II, pp. 649–654.

[22] Mitsoidis, D.A., N.A. Kampanis, and V.A. Dougalis (2008) Coupled mode and finite element approximations of underwater sound propagation problems in general stratified environments, *J. Comput. Acous.*, Vol. 16, pp. 83–116.

[23] Murphy, J.E. and S.A. Chin-Bing (1988) A finite element model for ocean acoustic propagation, *Math. Comput. Modelling*, Vol. 11, pp. 70–74.

[24] Murphy, J.E. and S.A. Chin-Bing (1989) A finite-element model for ocean acoustic propagation and scattering, *J. Acous. Soc. Am.*, Vol. 86, pp. 1478–1481.

[25] Antoine, X., H. Barucq, and A. Bendali (1999) Bayliss-Turkel-like radiation conditions on surfaces of arbitrary shape, *J. Math. Anal. Appl.*, Vol. 229, pp. 184–211.

[26] Bayliss, A., M. Gunzberger, and E. Turkel (1982) Boundary conditions for the numerical solution of elliptic equations in exterior domains, *SIAM J. Appl. Math.*, Vol. 42, pp. 430–451.

[27] Engquist, B. and A. Majda (1977) Absorbing boundary conditions for the numerical simulation of waves, *Math. Comput.*, Vol. 31, pp. 629–651.

[28] Keller, J.B. and D. Givoli (1989) Exact non-reflecting boundary conditions, *J. Comput. Phys*, Vol. 81, pp. 172–192.

[29] Givoli, D (1999). Recent advances in the DtN FE method, *Arch. Comput. Methods Eng.*, Vol. 6, pp. 71–116.

[30] Porter, M.B. and E.L. Reiss (1984) A numerical method for ocean-acoustic normal modes, *J. Acous. Soc. Am.*, Vol. 76, pp. 244–252.

[31] Fix, G.J. and S.P. Marin (1978) Variational methods for underwater acoustic problems, *J. Computat. Phys.*, Vol. 28, pp. 253–270.

[32] Vendhan C.P., G.C. Diwan, and S.K. Bhattacharyya (2010) Finite-element modeling of depth and range dependent acoustic propagation in ocean waveguides, *J. Acoust. Soc. Am.*, Vol. 127, pp. 3319–3326.

[33] Porter, M.B. and E.L. Reiss (1985) A numerical method for bottom interacting ocean acoustic normal modes, *J. Acous. Soc. Am.*, Vol. 77, pp. 1760–1767.

[34] Ihlenburg, F. (1998) *Finite Element Analysis of Acoustic Scattering*, Springer, Berlin.

[35] Taroudakis, M.I., G.A. Athanassoulis, and J.P. Ionnidis (1990) A variational principle for underwater acoustic propagation in a three-dimensional ocean environment, *J. Acoust. Soc. Am.*, Vol. 88, pp. 1515–1522.

[36] Bayliss, A., C.I. Goldstein, and E. Turkel (1985) The numerical solution of the Helmholtz equation for wave propagation problems in underwater acoustics, *Comput. Math. Appl.*, Vol. 11, pp. 655–665.

[37] Athanassoulis, G.A., K.A. Belibassakis, D.A. Mitsoudis, N.A. Kampanis, and V.A. Dougalis (2008) Coupled mode and finite element approximations of underwater sound propagation problems in general stratified environments, *J. Comput. Acoust.*, Vol. 16, pp. 83–116.

[38] Pekeris, C.L. (1948) Theory of propagation of explosive sound in shallow water, *Memoirs of Geological Society of America*, Vol. 27, pp. 1–116.

[39] Buckingham, M.J. and E.M. Giddens (2006) On the acoustic field in Pekeris waveguide with attenuation in the bottom half-space, J. Acous. Soc. Am., Vol. 119, pp. 124–142.

PERMISSIONS

All chapters in this book were first published in EARS, by InTech Open; hereby published with permission under the Creative Commons Attribution License or equivalent. Every chapter published in this book has been scrutinized by our experts. Their significance has been extensively debated. The topics covered herein carry significant findings which will fuel the growth of the discipline. They may even be implemented as practical applications or may be referred to as a beginning point for another development.

The contributors of this book come from diverse backgrounds, making this book a truly international effort. This book will bring forth new frontiers with its revolutionizing research information and detailed analysis of the nascent developments around the world.

We would like to thank all the contributing authors for lending their expertise to make the book truly unique. They have played a crucial role in the development of this book. Without their invaluable contributions this book wouldn't have been possible. They have made vital efforts to compile up to date information on the varied aspects of this subject to make this book a valuable addition to the collection of many professionals and students.

This book was conceptualized with the vision of imparting up-to-date information and advanced data in this field. To ensure the same, a matchless editorial board was set up. Every individual on the board went through rigorous rounds of assessment to prove their worth. After which they invested a large part of their time researching and compiling the most relevant data for our readers.

The editorial board has been involved in producing this book since its inception. They have spent rigorous hours researching and exploring the diverse topics which have resulted in the successful publishing of this book. They have passed on their knowledge of decades through this book. To expedite this challenging task, the publisher supported the team at every step. A small team of assistant editors was also appointed to further simplify the editing procedure and attain best results for the readers.

Apart from the editorial board, the designing team has also invested a significant amount of their time in understanding the subject and creating the most relevant covers. They scrutinized every image to scout for the most suitable representation of the subject and create an appropriate cover for the book.

The publishing team has been an ardent support to the editorial, designing and production team. Their endless efforts to recruit the best for this project, has resulted in the accomplishment of this book. They are a veteran in the field of academics and their pool of knowledge is as vast as their experience in printing. Their expertise and guidance has proved useful at every step. Their uncompromising quality standards have made this book an exceptional effort. Their encouragement from time to time has been an inspiration for everyone.

The publisher and the editorial board hope that this book will prove to be a valuable piece of knowledge for researchers, students, practitioners and scholars across the globe.

LIST OF CONTRIBUTORS

Pratima Pandey
Indian Institute of Remote Sensing, Dehradun, India

Alagappan Ramanathan
School of Environmental Science, Jawaharlal Nehru University, New Delhi, India

Gopalan Venkataraman
Centre of Studies in Resources Engineering, Indian Institute of Technology, Powai, Mumbai, India

Ana C. Teodoro
Earth Sciences Institute (ICT) and Department of Geosciences, Environment and Land Planning, Faculty of Sciences, University of Porto, Porto, Portugal

Dericks P. Shukla and Sharad Gupta
School of Engineering, Indian Institute of Technology, Mandi (HP), India

Chandra S. Dubey
Department of Geology, University of Delhi, Delhi, India

Manoj Thakur
School of Basic Sciences, Indian Institute of Technology, Mandi (HP), India

Marilia Ferreira Gomes and Philippe Maillard
Geography Department, Geosciences Institute, Federal University of Minas Gerais, Belo Horizonte, Brazil

Cartography Department, National Institute of Land Reform, Belo Horizonte, Brazil

Igor Ogashawara
Department of Earth Sciences, Indiana University – Purdue University at Indianapolis (IUPUI), Indianapolis, IN, USA

Marcelo P. Curtarelli, Carlos A. S. Araujo and José L. Stech
Remote Sensing Division, National Institute for Space Research (INPE), São José dos Campos, SP, Brazil

Monika Gähler
German Aerospace Center (DLR), Center for Satellite Based Crisis Information (ZKI) of the German Remote Sensing Data Center (DFD), Oberpfaffenhofen / Wessling, Germany

Saba Mudaliar
Sensors Directorate, Air Force Research Laboratory, WPAFB Ohio, USA

C.P. Vendhan
Indian Institute of Technology Madras, Chennai, India

C. Prabavathi
Independent researcher, Dayton, Ohio, USA

Index

www.ingramcontent.com/pod-product-compliance
Lightning Source LLC
Chambersburg PA
CBHW080401190526
45161CB00003B/103